THE SOCIETY FOR APPLIED BACTERIOLOGY
SYMPOSIUM SERIES NO. 4

MICROBIOLOGY IN AGRICULTURE, FISHERIES AND FOOD

Edited by

F. A. SKINNER

AND

J. G. CARR

1976

ACADEMIC PRESS

LONDON · NEW YORK · SAN FRANCISCO

A subsidiary of Harcourt Brace Jovanovich, Publishers

ACADEMIC PRESS INC. (LONDON) LTD
24-28 OVAL ROAD
LONDON N.W.1

U.S. Edition published by
ACADEMIC PRESS INC.
111 FIFTH AVENUE
NEW YORK, NEW YORK 10003

Library of Congress Catalog Card Number 75-45752
ISBN: 0-12-648060-5

Printed in Great Britain by
The Whitefriars Press Ltd., London and Tonbridge, England

MICROBIOLOGY IN
AGRICULTURE, FISHERIES
AND FOOD

Contributors

G. I. BARROW, *Public Health Laboratory, Royal Cornwall Hospital (City), Infirmary Hill, Truro TR1 2HZ, Cornwall, England*

MARGARET E. BROWN, *Soil Microbiology Dept., Rothamsted Experimental Station, Harpenden AL5 2JQ, Herts., England*

J. S. CROWTHER, *Unilever Research, Colworth House, Sharnbrook, Beds., England*

M. J. DAFT, *Dept. of Biological Sciences, The University, Dundee DD1 4HN, Scotland*

JENNIFER M. GEE, *ARC Food Research Institute, Colney Lane, Norwich NR4 7UA, England*

R. J. GILBERT, *Food Hygiene Laboratory, Central Public Health Laboratory, Colindale Avenue, London NW9 5HT, England*

A. HACKING, *A.D.A.S., Ministry of Agriculture, Fisheries & Food, Shardlow Hall, Shardlow, Derby, England*

J. HARRISON, *A.D.A.S., Ministry of Agriculture, Fisheries & Food, Shardlow Hall, Shardlow, Derby, England*

BETTY C. HOBBS, *Food Hygiene Laboratory, Central Public Health Laboratory, Colindale Avenue, London NW9 5HT, England*

P. N. HOBSON, *Rowett Research Institute, Bucksburn, Aberdeen, Scotland*

R. HOLBROOK, *Unilever Research, Colworth House, Sharnbrook, Beds., England*

D. E. HUGHES, *Corrosion & Protection Centre, The University of Manchester Institute of Science & Technology, P.O. Box 88, Manchester M60 1QD, England*

M. INGRAM, *ARC Meat Research Institute, Langford, Bristol BS18 7DY, England*

B. JARVIS, *British Food Manufacturing Industries Research Association, Randalls Road, Leatherhead, Surrey, England*

M. C. KAWAL, *Dept. of Food Technology, Iowa State University, Ames, Iowa 50010, U.S.A.*

A. A. KRAFT, *Dept. of Food Technology, Iowa State University, Ames, Iowa 50010, U.S.A.*

P. McDONALD, *Dept. of Agricultural Biochemistry, Edinburgh School of Agriculture, University of Edinburgh, West Mains Road, Edinburgh EH9 3JG, Scotland*

P. McKENZIE, *Corrosion & Protection Centre, The University of Manchester Institute of Science & Technology, P.O. Box 88, Manchester M60 1QD, England*

D. C. MILLER, *Public Health Laboratory, Royal Cornwall Hospital (City), Infirmary Hill, Truro TR1 2HZ, Cornwall, England*

N. J. MOON, *Dept. of Food Technology, Iowa State University, Ames, Iowa 50010, U.S.A.*

*J. L. OBLINGER, *Dept. of Food Technology, Iowa State University, Ames, Iowa 50010, U.S.A.*

J. L. PEEL, *ARC Food Research Institute, Colney Lane, Norwich NR4 7UA, England*

G. W. REINBOLD, *Dept. of Food Technology, Iowa State University, Ames, Iowa 50010, U.S.A.*

R. J. ROBERTS, *Unit of Aquatic Pathobiology, University of Stirling, Stirling FK9 4LA, Scotland*

F. A. SKINNER, *Soil Microbiology Dept., Rothamsted Experimental Station, Harpenden AL5 2JQ, Herts., England*

W. D. P. STEWART, *Dept. of Biological Sciences, The University, Dundee DD1 4HN, Scotland*

†A. J. TAYLOR, *Food Hygiene Laboratory, Central Public Health Laboratory, Colindale Avenue, London NW9 5HT, England*

H. W. WALKER, *Dept. of Food Technology, Iowa State University, Ames, Iowa 50010, U.S.A.*

H. WILLIAMS SMITH, *Houghton Poultry Research Station, Houghton, Huntingdon PE17 2DA, England*

* Present address: Department of Food Science, University of Florida, Gainsville, Florida 32601, U.S.A.
† Present address: Diagnostic Bacteriology Laboratory, St. Mary's Hospital Medical School (University of London), Norfolk Place, London W2, England

Preface

THE Society for Applied Bacteriology had its origin in dairy bacteriology and has always provided a forum for the discussion of research work and views on all aspects of the microbiology of agriculture, fisheries and food. In 1973 the Committee considered that it would be appropriate to devote the Symposium of the Summer Conference, due to be held at the University of Aberdeen in July 1974, to this subject. However, it was realized that the microbiology of our food production industries does not provide a topic sufficiently limited in scope for profitable discussion in the short space of two days. Contributors were asked, therefore, to discuss trends in research in their own fields, especially those that they considered to be important or controversial. The papers on diverse topics presented at that Symposium form the substance of this volume.

The Committee decided that the holding in Scotland of a Symposium on the Microbiology of Agriculture, Fisheries and Food would be a fitting occasion to pay tribute to the work of Dr Tom Gibson in agricultural microbiology and for the Society. Unfortunately, the plan to dedicate the Symposium to Dr Gibson was frustrated by his death in 1973. Instead, to mark the appreciation of the Society, it was decided to invite Professor M. Ingram, A.R.C. Meat Research Institute, Langford, Bristol, to deliver the Tom Gibson Memorial Lecture on a topic of his own choice. Professor Ingram's contribution takes pride of place in the following pages.

F. A. SKINNER
Rothamsted Experimental Station
Harpenden AL5 2JQ
Hertfordshire
England

J. G. CARR
University of Bristol
Research Station
Long Ashton
Bristol BS18 9AF
England

Contents

Mycotoxins in Food

B. JARVIS

The Tom Gibson Memorial Lecture

The Microbiological Role of Nitrite in Meat Products

M. INGRAM

*Meat Research Institute, Langford,
Bristol BS18 7DY, England*

CONTENTS

1. Introduction

INTEREST IN THIS subject has lately been greatly stimulated, by the demonstration that traces of carcinogenic nitrosamines may be formed in cured meats, through reaction of the nitrite used in curing with secondary and tertiary amines present in the meat. This has led to suggestions to restrict or forbid the use of nitrite, and also of nitrate which may serve as its precursor *(Status Report*, 1972).

On the other hand, besides providing the characteristic colour and flavour of cured meats, nitrite is believed to have important microbiological effects both on safety and stability (Leistner, 1973; Ingram, 1974). As regards safety, for example, there are still regular outbreaks of botulism caused by home-cured meats (Sebald, 1970; González & Guttiérrez, 1972) where nitrate and/or nitrite are either not used or ill controlled; whereas no botulism arises from the enormous volume of commercially cured meat where these salts are used in controlled amounts, despite the occasional demonstrable presence of *Clostridium botulinum* in such material (Abrahamsson & Riemann, 1971; T. A. Roberts, pers. comm.). Equally with stability, recent empirical investigations have shown a tendency to increased microbial growth when nitrite is diminished or omitted, both in cured meats which have been heated (Christiansen *et al.*, 1973, 1974) and those which have not (Shaw & Unsworth, 1973; Baird-Parker & Baillie, 1974).

We, therefore, ought to understand the microbiological action of nitrite; and, in particular, to seek a means of estimating how much nitrite is necessary in various circumstances, as a basis for possible restrictive legislation. It is better to consider unheated before heated systems, the latter being more complex: heating is, itself, an additional controlling factor; it adds chemical complications; and it makes the question one of spores rather than vegetative cells.

2. Unheated Systems

It was shown 40 years ago that nitrite by itself is relatively ineffective as an inhibitor of bacteria; and in curing practice it is always accompanied by salt and often nitrate, while normal meat is distinctly acid with a pH value near 6.0. Hence it has long been evident, in general terms, that the inhibitory effects depended on synergistic action between the various factors involved.

Indeed, interactions between pairs of factors were soon demonstrated. It was shown in 1940 with fish that the anti-microbial preservative effect of nitrite increases at lower pH values. This was later explained on the basis that it depends on the concentration of undissociated HNO_2; at the same time it was shown that the effect is greater under anaerobic than under aerobic conditions, and that some organisms (e.g. lactic acid bacteria) are relatively resistant (Castellani & Niven, 1955). Since 1947 I have emphasized that low pH increases the preservative effect of salt; the subject has recently been reviewed in terms of water activity, by Riemann, Lee & Genigeorgis (1972). So far as I know, it is only recently that a complementary action between the preservative effects of salt and nitrite has been recorded: first by Pivnick, Barnett, Nordin & Rubin (1969) referring to canned (i.e. heated) products, but the phenomenon has now been amply confirmed in unheated systems (e.g. Wood & Evans, 1973). In addition there have been demonstrated interactions of incubation temperature with limiting salt content (e.g. Segner, Schmidt & Boltz, 1966), with pH (Barnes, Despaul & Ingram, 1963), and with both (Baird-Parker & Freame, 1967; Alford & Palumbo, 1969).

(a) *Safety*

With food-poisoning bacteria, the combinations of pH, salt and nitrite have now been shown to have a triple interaction by Roberts & Ingram, (1973). Figure 1 gives a 3-dimensional presentation of the results with vegetative cells of types A, B, E, and F of *Cl. botulinum,* where each block represents a combination of circumstances which permitted growth from an inoculum of 1000 vegetative

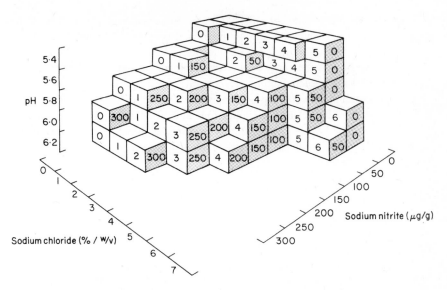

Fig. 1. The effect of pH, NaCl and NaNO$_2$ on growth from 10^3 vegetative cells of *Cl. botulinum* at 35°. (This diagram sums the data given separately for types A, B, E & F by Roberts & Ingram, 1973.)

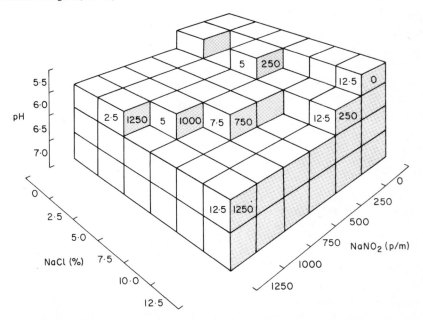

Fig. 2. The effect of pH, NaCl and NaNO$_2$ on growth from 10^4 undamaged cells of *Staph. aureus* (NCTC 10652) in heart infusion broth at 37° for 12 days. (Unpublished data of P. G. Bean & T. A. Roberts.)

cells/ml of one type or another, in 20 ml of culture medium. The surface of the pile of blocks thus represents the boundary between safe and unsafe combinations. It is at once evident that all 3 factors are equally important, and that nitrite cannot be considered by itself. The much greater opportunities of growth at higher pH values, such as may be deliberately induced for technical reasons (e.g. water-holding power), are strikingly evident. It appears that this triple inter-relation may be of general application. Similar though less detailed results have been obtained with *Cl. perfringens*; and Fig. 2 shows similar observations with *Staphylococcus aureus,* which is of course more resistant to salt save at low pH.

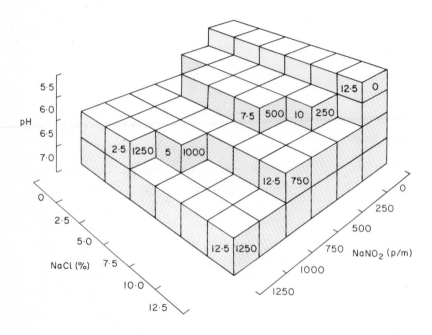

Fig. 3. The effect of pH, NaCl and NaNO$_2$ on growth from 4×10^3 undamaged cells of *Staph. aureus* (NCTC 10652) in heart infusion broth at 15° for 21 days. (Unpublished data of P. G. Bean & T. A. Roberts.)

Figure 3 illustrates the influence of incubation temperature. As comparison with Fig. 2 shows, many pH/salt/nitrite combinations which readily permit growth under warm conditions, are wholly inhibitory at lower temperature. Experiments now in progress show that the same holds for the clostridia. The requirement to store semi-conserves under cool conditions must have been largely based, unknowingly, on these phenomena.

There is a fifth important factor, yet to be satisfactorily dealt with, i.e. the number of cells involved. In general, the smaller the number, the smaller the inhibitory concentrations needed; a relation previously known with spores but not with vegetative cells. With vegetative cells, there is little difference until inocula are reduced to *c.* 100 cells, in volumes of 20 ml; at lower levels than this, weaker combinations suffice to inhibit.

We can imagine how it might become possible to evaluate the effects of temperature and cell number. Figure 4 shows an illustration by Roberts (1974) on the basis that the nitrite/pH/salt boundary surface is part of a sphere. If the temperature is lower, or the number of cells smaller, the boundary conditions are lower, i.e. they are contained in a smaller sphere. Temperature and cell number can, thus, be allowed for by adjusting the radius of the sphere. Whether the boundary surface should be plane, spherical, or ellipsoidal, as hinted in Fig. 4, remains to be established. Also, we still lack satisfactory mathematical means to express the influences of cell numbers and temperature, and we have yet to explore the mutual interaction of these 2 factors.

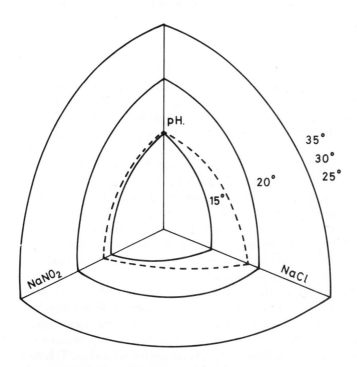

Fig. 4. Schematic illustration of the effect of incubation temperature on inhibition of *Cl. botulinum* by pH, NaCl and $NaNO_2$. (Roberts, 1974.)

Besides the foregoing 5 interacting factors, there are several others which might be involved. Nitrate is commonly assumed to have no distinct anti-microbial role in unheated products (though it can serve as a reserve precursor of nitrite, and trivially assist salt in reducing the water activity); the point has not been carefully investigated, however, hence nitrate has been omitted from the present account. On slender evidence, we believe that glucono-delta-lactone, alkali and phosphates have no effect other than through changing the pH.

The possible role of ascorbic acid is of special interest, for it has been shown greatly to diminish the formation of nitrosamines in cured meat; it is, indeed, regarded as a general anti-nitrosating agent, because of its ability to react with nitrite especially in acid conditions (N.B. in estimating nitrite in presence of ascorbate!). The question is, would ascorbic acid similarly diminish the desirable anti-microbial effects of nitrite? In meats where the nitrite is not heated, it has recently been indicated that ascorbic acid has little effect (Baird-Parker & Baillie, 1974; their Tables 11 & 12); which seems surprising, but the data published do not reveal whether the expected decrease of nitrite concentration actually occurred. At least it appears justifiable, for the present, to concentrate on the inter-relations between the 5 principal factors already discussed.

(b) *Stability*

Let us now glance at the stability of unheated cured meats. It is well known that properly cured meats are much less perishable than fresh, while they do not undergo a putrid type of spoilage and support a flora consisting mainly of micrococci. This was previously ascribed to the ability of many *Micrococcus* strains to tolerate salt like staphylococci, and no particular role for nitrite was envisaged.

Recently, however, there have been clear indications that nitrite does play a part in establishing the flora characteristic of cured meat (Wood & Evans, 1973). A distinct contribution might perhaps be expected, for organisms which resist salt are not necessarily those which resist nitrite — e.g. lactic acid bacteria are more resistant to nitrite, and less to salt, than are micrococci and staphylococci. If the observations with pathogens are any guide, complex synergistic actions with salt, pH etc. are also likely to occur in this connection. Much more detailed information on these topics is needed.

Further, some of the salt-tolerant bacteria are proteolytic, hence it seems possible that the absence of putrefactive odours in the spoilage of cured meat might be due to an inhibition by nitrite (as previously postulated for salt by Ingram & Kitchell, 1968) of the relevant biochemical processes notably the decarboxylation of amino acids to produce offensive amines. This possibility has recently been explored by Dainty & Meredith (1973), who failed to find any such effect separate from that on growth.

3. Heated Systems

Because vegetative cells are destroyed by heating, interest in heated systems is specially related to inhibitory effects on bacterial spores, especially those of *Cl. botulinum* and of the more heat-resistant 'indicator' strains of *Cl. sporogenes* like PA 3679. The topic has been extensively investigated to try to explain the well established but ill explained fact that canned cured meats are safe and stable with only a fraction of the heat process needed for un-cured meats. In addition, there are salt- and nitrite-resistant *Bacillus* strains, which form gas (N_2, N_2O) from nitrate or nitrite and are important in spoiling pasteurized cured meats (Eddy & Ingram, 1956).

A series of inoculated pack experiments, mainly in the U.S.A., established the need for salt and nitrite, with the heating, and showed that combinations which inhibited small numbers of spores failed to control larger numbers (reviewed by Spencer, 1966). They culminated in the statistically based experiment by Riemann (1963), illustrated in Table 1. This confirmed the importance of interactions between heat, salt and nitrite. It is noteworthy that nitrate also was of significance; and that no interaction was revealed between pH and nitrite. Recently, inoculated pack experiments have been resumed in the U.S.A. on a large scale, repeatedly demonstrating a general relation between the concentration of nitrite and inhibition of inoculated *Cl. botulinum* (Greenberg, 1972; Christiansen *et al.*, 1973; Hustad *et al*, 1973; Christiansen *et al.*, 1974).

Table 1
*Factors affecting stability of canned cured meats
(Riemann, 1963)*

Factors tested	NaCl; $NaNO_3$; $NaNO_2$; pH; F_0
Significant singly	All except $NaNO_3$
Significant interactions	$F_0 \times NaCl$
	$F_0 \times NaNO_2$
	$NaCl \times NaNO_3 \times NaNO_2$
	$NaCl \times pH$
	$NaCl \times NaNO_3$

The disadvantage of experiments of this kind is, however, that it is difficult to see what takes place inside the pack. Thus we cannot decide between 4 possibilities, first clearly enunciated by Johnston, Pivnick & Samson (1969): (1) whether the heat resistance of the spore itself is lowered (2) whether there is increased germination followed by death of the germinated spores (3) whether spores which have been heated are more readily inhibited (4) whether heating produces some extra inhibitory substance. Knowledge of these possibilities is needed to develop a rational interpretation of the phenomena.

(a) *The enhanced destruction of spores by heat*

This possibility, suggested by Jensen & Hess (1941), is now largely eliminated. The effect of heating in the presence of realistic concentrations is small, so far as sodium chloride (Roberts, Gilbert & Ingram, 1966) and nitrate (Duncan & Foster, 1968*a*) are concerned. As regards nitrite, results are ambiguous: Duncan & Foster (1968*a*) claimed that the presence of nitrite during heating increased heat sensitivity (though the data are confusing); whereas Ingram & Roberts (1971) observed no corresponding phenomenon.

(b) *Germination during the heating process*

This was soon negatived, as a general explanation, by observations showing that many spores are in fact not killed (e.g. Silliker, Greenberg & Schack, 1958; Riemann, 1963). Indeed, Silliker, Greenburg & Schack (1958) calculated that 20% of their inoculated spores survived in heated cured meat packs which nevertheless remained stable; Pivnick & Chang (1974) made similar observations recently.

(c) *Extra inhibitory effect of salts consequent on heat damage*

Silliker *et al.* (1958) therefore speculated that the heated spores must be inhibited by the curing salts. Aided first by my deceased colleague B. P. Eddy and then by T. A. Roberts, I examined this possibility in model experiments. Large bacteriostatic effects were observed when spores, heated in whatever

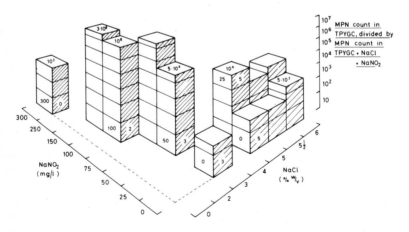

Fig. 5. Combined effects of NaCl and unheated $NaNO_2$ in inhibiting *Cl. botulinum* spores (in medium of Roberts & Ingram, 1973, plus 0.4% glucose added separately) at pH 6.2 and 37°. (Roberts, 1974.)

system, were plated out on media containing salt or nitrite (Roberts & Ingram, 1966); a result soon confirmed by Duncan & Foster (1968a). With nitrite, this was especially so at acid pH values, and Roberts & Ingram (1966) remarked that the relations suggested the involvement of undissociated HNO_2.

In addition there are indications, illustrated in Fig. 5, of a synergistic effect of salt with nitrite; and, from the observations of Baird-Parker & Baillie (1974) (their Table 1) it seems likely that there is similarly a synergistic effect of incubation temperature.

It will be observed that, in Fig. 5, the data are presented in terms of decimal reductions brought about by particular treatments (cf. Pivnick & Chang, 1974). This form of presentation has potentially a great advantage in considering heated cured meats. According to our observations, the effect of heat treatment on number of viable spores may be evaluated on a similar basis (D value) even in cured meats (Fig. 6). Therefore, presentation in terms of decimal reductions should permit the effects of salts to be compared and (hopefully) combined with those of heat treatment and number of spores.

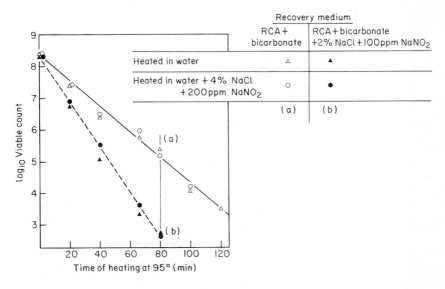

Fig. 6. The effect of sodium chloride and sodium nitrite on heat resistance and recovery of spores of *Cl. botulinum* 33A (Ingram & Roberts, 1971); illustrating the applicability of a decimal reduction (D value) concept in presence as in absence of curing salts.

This, however, is only part of the story in heated systems, for it refers to nitrite as such, and we still have to consider the possible effect of derived substances produced by heating.

(d) *Production of more inhibitory substances from nitrite*

Both Castellani & Niven (1955) and Gough & Alford (1965) observed that sterilization of nitrite plus glucose reduced the concentration necessary to inhibit, the former authors ascribing this to the production of more anaerobic conditions. Perigo, Whiting & Bashford (1967) observed that the inhibitory concentrations of nitrite in the experiments of Roberts & Ingram (1966) were of the order 100–500 mg/kg, some 10-fold greater than the concentrations effective in canning practice; and they noted that Roberts & Ingram (1966) had added filter-sterilized cold nitrite solution, instead of heating nitrite with the supporting medium as in canning.

(i) *Culture media*

Perigo *et al.* (1967) then showed, with a suitable culture medium, that about 10 times more nitrite was needed to inhibit, if it were added after instead of before autoclaving; hence they concluded that the nitrite was converted, by heating in the medium, into a substance (or substances) at least 10 times more inhibitory. Their observation has been thoroughly confirmed in several investigations (e.g. Table 2), using culture media. The active agent is referred to as the Perigo Factor.

Table 2

Inhibitory concentrations (mg/kg) of sodium nitrite and heated sodium nitrite (pH 6) (Roberts & Smart, 1974)

Spores		Unheated nitrite			Heated nitrite		
Treatment	No./ml	A	E	S	A	E	S
Unheated	10^6	240	160	200	20	15	10
	10^3	240	240	200	15	7.5	10
Heated	10^6	200	160	200	40	5	15
	10^3	200	240	200	10	5	15
Irradiated	10^6	240+	240+	240+	40	15	20
	10^3	240	200	200	40	7.5	20

A, *Cl. botulinum* Type A; E, *Cl. botulinum* Type E; S, *Cl. sporogenes*.

It has since been suggested that this inhibition is due to Roussin salts, co-ordination complexes containing the group $-S-Fe(NO)_2$, formed when nitrite is heated with $-SH$ compounds in presence of ferrous ions. When artificial solutions were prepared in this way with H_2S or cysteine, the solutions were inhibitory where the unheated mixtures were not (van Roon, 1974); and the black salt became partially inhibitory to *Cl. sporogenes* spores at concentrations of *c.* 15 mg/l and completely so at about twice that level. But this, while suggestive, scarcely proves that Roussin salts cause the inhibition even in culture media, still less so in meat.

The inhibitory action of the Perigo Factor was shown by its discoverers to be little affected by change of pH, unlike that of nitrite itself; and this agrees with Riemann's (1963) findings, that pH is relatively unimportant in heated cured meat. On the other hand, there are several reasons for doubting that the Perigo Factor of culture media is involved in meat. For example, unlike the case in meat, in culture media heating at temperatures $> 90°$, preferably $c.$ $110°$, is necessary (Perigo *et al.*, 1967); and the number of spores, and heating or irradiation of the spores beforehand, make little difference to the inhibitory concentrations (cf. Table 2). Attempts to detect Roussin salts in heated cured meat have so far failed (van Roon, 1974); and it was shown by Johnston, Pivnick & Samson (1969), and confirmed by others, that if a small proportion of lean meat is added to a culture medium already containing Perigo Factor, its inhibitory activity disappears. Observations by Grever (1974) suggest that it is destroyed by (oxidized?) fat, and that it is adsorbed and inactivated by meat fibres. Johnston *et al.* (1969) and Johnston & Loynes (1971) showed that the Perigo Factor in media is not dialysable, whereas the inhibitor in a meat system is.

It seems that observations made on culture media may have little relevance to meat; for which reason it is better, for the time being, not to use the terms Perigo Factor or Inhibitor in connection with meat: much work with nitrite heated in media is, therefore, omitted from this account.

(ii) *Cured Meat*

It proves, in fact, that similar phenomena are much less obvious in meat systems. The first demonstration (Table 3) was by Ashworth & Spencer (1972),

Table 3

A so-called Perigo Effect in pork (data of Ashworth & Spencer, 1972)

NaNO$_2$ added (mg/kg)	Residual NaNO$_2$‡ (mg/kg) After*	Residual NaNO$_2$‡ (mg/kg) Before†	Growth in No. of replicates (of 4) after incubation at 37° for (days) After* 1	3	10	Before† 1	3	10
0	0	0	4	4	4	4	4	4
75	52	–	4	4	4	–	–	–
100	68	–	2	4	4	–	–	–
150	114	32	0	1	3	2	4	4
200	142	36	0	0	0	1	4	4
300	219	66	0	0	0	0	2	2
400	–	108	–	–	–	0	0	0

Nitrite added before (†) or after (*) heating the meat.
‡ At time when bacteria added, i.e. time of challenge.

who found that the effect increased with the amount of heating, and that pH had little influence. Confirmatory observations were made by Ashworth, Hargreaves & Jarvis (1973), who found the effect so variable that statistical treatment was necessary to confirm its existence; sometimes the effect was negative in terms of the 'input' concentrations added initially (e.g. Table 4), and it was always relatively small. However, on the basis of residual nitrite concentrations (at time of challenge!) they detected a synergistic action of salt (Table 4) and an effect even at pasteurization temperatures (Table 5).

These experiments demonstrate a 'Perigo Effect', in the original sense that less input nitrite was usually needed if added before instead of after heating; but they do nothing to demonstrate a connection with the Perigo Factor of culture media. Johnston & Loynes (1971) failed to find any non-dialysable inhibitor in heated meat + nitrite, even with added reducing agents. Nevertheless, the effective inhibitor in the meat system may not be nitrite; for Johnston & Loynes (1971), while observing that reducing agents much increased the inhibitory

Table 4

Effect of sodium chloride on ED_{100} of nitrite (p/m $NaNO_2$) against Cl. sporogenes *(Ashworth et al. 1973)*

	Nitrite added before heat		Nitrite added after heat	
NaCl	Input	Residual	Input	Residual
0	200–500	210–220	150–300	106–336
3.5%	75–250	32–104	100–300	64–343

Table 5

ED_{100} *values of residual nitrite (p/m $NaNO_2$) against* Cl. sporogenes *(Ashworth et al. 1973)*

	80° 4 h	20–70° in 4 h	115° 20 min
Vegetative cells			
Before*	122	109	64
After†	152	185	91
Spores (added before heat)			
Before	84	149	–
After	194	222	–
Spores (added after heat)			
Before	84	102	–
After	123	234	–

* Before, nitrite added before heating.
† After, nitrite added after heating.

action of nitrite (their Fig. 1), give data showing that those agents increased the disappearance of nitrite in almost the same proportions (their Fig. 2). This suggests that the nitrite which disappeared was mainly responsible for the inhibition in their meat system.

The clearest indication of a nitrite-derived inhibitor in meat has come recently from an ingenious experiment by Pivnick & Chang (1974). They cured meat with different amounts of nitrite, canned following normal commercial procedure, and then stored the cans until all the nitrite had disappeared; after which, they inoculated with spores appropriately 'damaged' by heating in a curing solution. They observed a period of inhibition which was longer the greater the input concentration of nitrite (their Table 6). This plainly suggests that the nitrite had been converted into some inhibitor. To get long periods of stability, however, they had to begin with unusually high concentrations of nitrite, which might have been expected from the unusual experimental conditions.

Ashworth et al. (1973) remarked that, in presence of salt, the inhibitory nitrite concentrations they observed were below the levels at present permitted in cured meats; but those concentrations, perhaps owing to use of relatively high incubation temperature $(37°)$ and spore numbers $(20/g)$, may nevertheless be substantially higher than those used in actual practice.

(e) Outstanding questions

Such observations, however, leave major questions still outstanding.

One is, how may experimental findings be extrapolated to estimate the probability of inhibition in a realistic quantity of material? Ashworth & Spencer (1972) calculated ED_{50} nitrite concentrations from 4 containers of 25 g containing unstated, but probably large, numbers of cells or spores; Ashworth et al. (1973) even calculated ED_{100} values, from the same number of containers each with 500 spores totalling 2000 spores; Christiansen et al. (1973) used 40 replicates with c. 700 spores each to infer that 200 p/m of nitrite gave 'zero probability of toxicity'. Had these workers used larger amounts of material and organisms, it seems certain that greater quantities of nitrite would have been required to ensure inhibition; as in the trials of Christiansen et al. (1973, 1974) who added more spores per container. How much nitrite would be needed to ensure inhibition over, say, 10^7 containers of 100 g containing 1 spore/20 containers − i.e. 500 000 spores treated individually? We do not yet know how to approach this question where nitrite and salt are involved, whether in heated or unheated systems.

A second question is, how much of the inhibition in a meat system is due to nitrite itself, and how much to derived inhibitor? We do not know how to handle this question either. The answer may depend on the nature of the meat, accounting for the observed variability in results: it is noteworthy that, in contrast to the large disappearance of nitrite on heating in the experiments of Ashworth & Spencer (1972), Christiansen *et al.* (1973) found only small changes on cooking (i.e. autoclaving).

This relates to a further question: since the nitrite concentration falls from the moment of its addition, through heating and storage, at what point should it be measured to give the best index of inhibition? Christiansen *et al.* (1973) used the initial ('input') concentration; Greenberg (1972) states, without demonstrating, that input concentration relates to inhibition better than does residual concentration; input concentration was the index in the experiments of Pivnick & Chang (1974) even when all the nitrite had disappeared as such. On the other hand, Ashworth & Spencer (1972) and Ashworth *et al.* (1973) emphasize the residual nitrite at the time of challenge (i.e. inoculation). In fact, some data of Ashworth & Spencer (1972) (their Table 7) show that much larger nitrite concentrations at time of challenge are required to secure inhibition of dormant spores over longer periods of storage, plainly suggesting that the residual level after a period of storage is critical. Until it is clear whether the inhibitory effect at different times is mainly due to nitrite as such, or to a derived inhibitor, or to both, we cannot decide whether the best critical index is the initial nitrite concentration, or the residual, or the amount of nitrite which has disappeared, or some uncertain fraction of the latter. In this position, it would obviously be difficult to produce satisfactory quantitative regulations.

4. The Mechanism of Action

The alternative, to find an adequate substitute for nitrite, might be easier if we knew how it acts. The previous investigations showed that in unheated systems the desirable effects of nitrite depend on its ability to inhibit growth of the relevant vegetative bacteria. As regards the spores which are critical in heated systems, Duncan & Foster (1968*b*) showed that, used alone, 4% $NaNO_2$ did not prevent germination of clostridium spores, while 0.06% at pH 6 or 0.8% at pH 7 stopped outgrowth, and less sufficed to prevent subsequent cell division. It therefore appears now that, in heated or in unheated systems, the critical effect of nitrite depends on its ability to prevent vegetative growth; though it should be confirmed that the same holds for its effect in synergistic combinations. But, apart from the relation with pH and the implication that conversion to HNO_2 is an essential first step, these investigations do nothing to elucidate the nature of the chemical mechanism involved. As already inferred, the nature and effect of the heated nitrite inhibitor are still unknown.

It has been assumed that nitrite blocks energy metabolism by reacting with the amino groups of dehydrogenases; but recent investigations by Dainty & Meredith (1973) suggest a different explanation, while confirming that nitrite can inhibit various respiratory enzymes. They found that the limiting concentration of nitrite which inhibits growth and synthetic processes is the same as that which inhibits uptake of substrates for macromolecular synthesis, and it also inhibits uptake of glucose and exogenous glucose respiration, whereas concentrations several-fold greater are needed to inhibit endogenous respiration; similar results were obtained with dissimilar species (*Ps. aeruginosa, Staph. aureus, Microb. thermosphactum*), though at different limiting concentrations. It is, thus, likely that the growth-inhibitory effect of nitrite is because it generally inhibits the uptake of energy sources.

There are various other substances which might do the same, but few are likely to be acceptable as food additives; and, it must be remembered, nitrite has other essential functions besides the microbiological.

5. Conclusions

The foregoing considerations can be summed up as follows.

In unheated systems, the effect of nitrite depends on 4 similarly significant factors: salt, pH, incubation temperature, also the number of cells involved. In principle, we can now see how to evaluate the interaction of these 5 factors. We need more detailed information about the relations of the last 2; and it should be confirmed that the influence of nitrate and ascorbic acid is negligible.

Systems involving heating present a more complex situation. Spores damaged by heating are more readily inhibited by nitrite, and there are some indications that the action of pH and salt is synergistic; the influence of spore number is already known; that of incubation temperature is not. But these relations refer to spores heated to temperatures of $c.$ $100°$, and it is still doubtful how far they apply to pasteurization temperatures of $c.$ $70°$. Further, experiments in culture media have shown that when nitrite is heated in the medium it may be converted into a more powerful inhibitor, though it is doubtful if the same inhibitor is produced in meat. Nevertheless, recent experiments have indicated that some inhibitor is occasionally produced when nitrite is heated in meat, even at temperatures as low as $70°$. But the quantitative importance of this factor remains to be established; and, until it has been we do not know whether the best index of the inhibitory effect of nitrite is the initial or residual concentration, or some fraction of the amount of nitrite which has disappeared. There are some indications that this heated meat + nitrite inhibitor is stimulated by salt, and not affected by pH, but quantitative elucidation of such influences awaits a procedure for separating the inhibition due to inhibitor from that due to nitrite itself.

The prospects of finding a satisfactory substitute for nitrite do not appear promising, even if attention be confined to its microbiological effects and its other technologically important properties are ignored. Consequently it has seemed more useful, for the present, to aim for a quantitative understanding of its microbiological effects, in order to provide a rational basis for any regulation of permissible quantities.

6. References

ABRAHAMSSON, K. & RIEMANN, H. (1971). Prevalence of *Clostridium botulinum* in semi-preserved meat products. *Appl. Microbiol.* **21**, 543.

ALFORD, J. A. & PALUMBO, S. A. (1969). Interaction of salt, pH, and temperature on the growth and survival of salmonellae in ground pork. *Appl. Microbiol.* **17**, 528.

ASHWORTH, J. & SPENCER, R. (1972). The Perigo effect in pork. *J. Fd Technol.* **7**, 111.

ASHWORTH, J., HARGREAVES, L. L. & JARVIS, B. (1973). The production of an antimicrobial effect in pork heated with sodium nitrite under simulated commercial pasteurisation conditions. *J. Fd Technol.* **8**, 477.

BAIRD-PARKER, A. C. & BAILLIE, M. A. H. (1974). The inhibition of *Cl. botulinum* by nitrite and sodium chloride. Eds B. Krol & B. J. Tinbergen. *Proc. int. Symp. Nitrite Meat Prod., Zeist, 1973*, p. 77 (Pudoc, Wageningen).

BAIRD-PARKER, A. C. & FREAME, B. (1967). Combined effect of water activity, pH and temperature on the growth of *Cl. botulinum* from spore and vegetative cell inocula. *J. appl. Bact.* **30**, 420.

BARNES, E. M., DESPAUL, J. E. & INGRAM, M. (1963). The behaviour of a food poisoning strain of *Cl. welchii* in beef. *J. appl. Bact.* **26**, 415.

CASTELLANI, A. G. & NIVEN, C. F. Jr. (1955). Factors affecting the bacteriostatic action of sodium nitrite. *Appl. Microbiol.* **3**, 154.

CHRISTIANSEN, L. N., JOHNSTON, R. W., KAUTTER, D. A., HOWARD, J. W. & AUNAN, W. J. (1973). Effect of nitrite and nitrate on toxin production by *Cl. botulinum* and on nitrosamine formation in perishable canned comminuted cured meat. *Appl. Microbiol.* **25**, 357.

CHRISTIANSEN, L. N., TOMPKIN, R. B., SHAPARIS, A. B., KUEPER, T. V., JOHNSTON, R. W., KAUTTER, D. A. & KOLARI, O. J. (1974). Effect of sodium nitrite on toxin production by *Cl. botulinum* in bacon. *Appl. Microbiol.* **27**, 733.

DAINTY, R. H. & MEREDITH, G. C. (1973). Mechanisms of inhibition of growth of bacteria by nitrite. *Ann. Rept. Meat Res. Inst.* 1972-73, p. 82. London: H.M.S.O.

DUNCAN, C. L. & FOSTER, E. M. (1968*a*). Role of curing agents in the preservation of shelf stable canned meat products. *Appl. Microbiol.* **16**, 401.

DUNCAN, C. L. & FOSTER, E. M. (1968*b*). Effect of sodium nitrite, sodium chloride, and sodium nitrate on germination and outgrowth of anaerobic spores. *Appl. Microbiol.* **16**, 406.

EDDY, B. P. & INGRAM, M. (1956). A salt-tolerant, denitrifying *Bacillus* strain which 'blows' canned bacon. *J. appl. Bact.* **19**, 62.

GONZÁLEZ, C. & GUTIÉRREZ, C. (1972). Intoxication botulique humain par *Cl. botulinum* B. *Annls Inst. Pasteur, Paris* **123**, 799.

GOUGH, B. J. & ALFORD, J. A. (1965). Effects of curing agents on the growth and survival of food-poisoning strains of *Clostridium perfringens*. *J. Fd Sci.* **30**, 1025.

GREENBERG, R. A. (1972). Nitrite in the control of *Cl. botulinum*. *Proc. American Meat Inst. Found. Meat Ind. Res. Conf., March 1972*, p. 25.

GREVER, A. B. G. (1974). Discussion. Eds B. Krol & B. J. Tinbergen. *Proc. int. Symp. Nitrite Meat Prod., Zeist, 1973*, p. 122 (Pudoc, Wageningen).

HUSTAD, G. O., CERVENY, J. G., TRENK, H., DEIBEL, R. H., KAUTTER, D. A., FAZIO, T., JOHNSTON, R. W. & KOLARI, O. E. (1973). Effect of sodium nitrite and sodium nitrate on botulinal toxin production and nitrosamine formation in wieners. *Appl. Microbiol.* **26**, 22.

INGRAM, M. (1974). The microbiological effects of nitrite. Eds B. Krol & B. J. Tinbergen. *Proc. int. Symp. Nitrite Meat Prod., Zeist, 1973*, p. 63 (Pudoc, Wageningen).

INGRAM, M. & KITCHELL, A. G. (1968). Salt as a preservative. *J. Fd Technol.* **3**, 77.

INGRAM, M. & ROBERTS, T. A. (1971). Application of the D-concept to heat treatments involving curing salts. *J. Fd Technol.* **6**, 21.

JENSEN, L. B. & HESS, W. R. (1941). A study of the effects of sodium nitrate on bacteria in meat. *Fd Mf.* **16**, 157.

JOHNSTON, M. A. & LOYNES, R. (1971). Inhibition of *Clostridium botulinum* by sodium nitrite as affected by bacteriological media and meat suspensions. *Can. Inst. Fd Technol. J.* **4**, 179.

JOHNSTON, M. A., PIVNICK, H. & SAMSON, J. M. (1969). Inhibition of *Clostridium botulinum* by sodium nitrite in a bacteriological medium and in meat. *Can. Inst. Fd Technol. J.* **2**, 52.

LEISTNER, L. (1973). Welche Konsequenzen hätte ein Verbot oder eine Reduzierung des Zusatzes von Nitrat und Nitritpökelsalz zu Fleischerzeugnissen? *Die Fleischwirtschaft* **53**, 375 & 378.

PERIGO, J. A., WHITING, E. & BASHFORD, T. E. (1967). Observations on the inhibition of vegetative cells of *Clostridium sporogenes* by nitrite which has been autoclaved in a laboratory medium, discussed in the context of sub-lethally processed meats. *J. Fd Technol.* **2**, 377.

PIVNICK, H., BARNETT, H. W., NORDIN, H. R. & RUBIN, L. J. (1969). Factors affecting the safety of canned, cured, shelf-stable luncheon meat inoculated with *Clostridium botulinum. Can. Inst. Fd Technol. J.* **2**, 141.

PIVNICK, H. & CHANG, P–C. (1974). Perigo effect in pork. Eds B. Krol & B. J. Tinbergen. *Proc. int. Symp. Nitrate Meat Prod., Zeist, 1973*, p. 111. (Pudoc, Wageningen).

RIEMANN, H. (1963). Safe heat processing of canned cured meats with regard to bacterial spores. *Fd Technol., Champaign*, **17**, 39.

RIEMANN, H., LEE, W. H. & GENIGEORGIS, C. (1972). Control of *Clostridium botulinum* and *Staphylococcus aureus* in semi-preserved meat products. *J. Milk Fd Technol.* **35**, 514.

ROBERTS, T. A. (1974). Inhibition of bacterial growth in model systems in relation to the stability and safety of cured meats. Eds B. Krol & B. J. Tinbergen. *Proc. int. Symp. Nitrite Meat Prod., Zeist, 1973*, p. 91. (Pudoc, Wageningen).

ROBERTS, T. A., GILBERT, R. J. & INGRAM, M. (1966). The effect of sodium chloride on heat resistance and recovery of heated spores of *Cl. sporogenes* (PA3679/S_2). *J. appl. Bact.* **29**, 549.

ROBERTS, T. A. & INGRAM, M. (1966). The effect of sodium chloride, potassium nitrate and sodium nitrite on recovery of heated bacterial spores. *J. Fd Technol.* **1**, 147.

ROBERTS, T. A. & INGRAM, M. (1973). Inhibition of growth of *Cl. botulinum* at different pH values by sodium chloride and sodium nitrite. *J. Fd Technol.* **8**, 467.

ROBERTS, T. A. & SMART, J. L. (1974). Inhibition of spores of *Clostridium* spp. by sodium nitrite. *J. appl. Bact.* **37**, 261.

SEBALD, M. (1970). Sur le botulism en France de 1956 à 1970. *Bull. Acad. Nat. Méd.* **154**, 703.

SEGNER, W. P., SCHMIDT, C. P. & BOLTZ, J. K. (1966). The effect of sodium chloride and pH on the outgrowth of spores of type E *Clostridium botulinum* at optimal and sub-optimal temperatures. *Appl. Microbiol.* **14**, 49.

SHAW, B. G. & UNSWORTH, B. (1973). Microbiology of the curing of bacon in the absence of nitrate and with diminishing quantities of nitrite. *Ann. Rept. Meat Res. Inst.* 1972-73, p. 77. London: H.M.S.O.

SILLIKER, J. H., GREENBERG, R. A. & SCHACK, W. R. (1958). Effect of individual curing ingredients on the shelf stability of canned comminuted meats. *Fd Technol., Champaign* **12**, 551.

SPENCER, R. (1966). Processing factors affecting stability and safety of non-sterile canned cured meats. *Fd Mf.* **41**, 39.

STATUS REPORT. (1972). Nitrites, nitrates, and nitrosamines in food – a dilemma. *J. Fd Sci.* **37**, 989.

VAN ROON, P. S. (1974). Inhibitors in cooked meat products. Eds B. Krol & B. J. Tinbergen. *Proc. int. Symp. Nitrite Meat Prod., Zeist, 1973,* p. 117. (Pudoc, Wageningen).

WOOD, J. M. & EVANS, G. C. (1973). Curing limits for mild cured bacon. *Proc. Inst. Fd Sci. Technol.* **6,** 111.

Methodology in Soil Examination

F. A. SKINNER

Soil Microbiology Department, Rothamsted Experimental Station,
Harpenden, Hertfordshire, England

CONTENTS

1. Introduction

A SURVEY of the literature shows that microbiologists understand clearly how the advance of their science depends on the continual development of new or improved methods: the soil microbiologist is not exceptional in his attitude. For example, at a recent meeting on 'Modern Methods in the Study of Microbial Ecology' at Uppsala (Rosswall, 1973) some 40 papers dealt with methods for the study of micro-organisms in soil. Against such a background the presentation of one paper on methods seems at first sight to be presumptuous and futile, but it may not be so if attention is focussed on a few problems of fundamental importance in soil microbiology and on the methods that have been applied to study them.

Soil microbiology has passed through several main phases of development though they are not sharply separated from each other. In the first, exploratory period soil was examined to see which micro-organisms were present and what they did. The methods of isolation and characterization of micro-organisms were, for the most part, those used by medical bacteriologists of the time. These studies revealed the presence of many micro-organisms of diverse morphological types and unusual properties, and indicated the microbial nature of several soil processes important to agriculture; examples are ammonification, nitrification, nitrogen fixation and denitrification. During this time much was learned about the decomposition of naturally occurring organic compounds such as cellulose, and about the association of micro-organisms with higher plants, for example, *Rhizobium* with the roots of legumes.

Throughout this period there was an increasing awareness that many micro-organisms, when added to soil, were inhibited or destroyed by other organisms already living there (see Waksman & Woodruff, 1940). The realization that microbial antagonists often acted by producing inhibitory substances, some of which, like penicillin, were clearly of medical importance, led to a second phase in the development of soil microbiology. This period, which immediately followed World War II, was one of exploitation in which the soil was regarded by many as a source of micro-organisms with useful industrial potential. The discovery of streptomycin (Schatz, Bugie & Waksman, 1940) focussed attention on the actinomycetes and stimulated a world-wide, purposeful search for those able to produce antibiotics of clinical value. This search was remarkably successful. Interesting offshoots of this research were detailed studies on taxonomy and classification of the actinomycetes (see Cross & Goodfellow, 1973), and the realization that antagonism between micro-organisms in soil, resulting from the production of inhibitory substances, might be an important factor in regulating the growth of soil microbial populations.

The clear relationships between soil microbial activity and the production of nutrients for higher plants has always made the soil worker feel that soil fertility might be at least partly explainable in terms of the behaviour of microbial populations. However, attempts to correlate fertility, usually assessed by crop yield, with the size or activity of the soil microbial population have met with little success; evidently we have not asked the right questions of nature. Nevertheless, the realization that such a connection exists does, I suspect, explain the enormous current interest in soil microbial ecology.

A deep understanding of the true nature and behaviour of the soil microflora is essential before the soil microbiologist can make an effective contribution towards solving the basic problem of agriculture, the maintenance of soil fertility. We are now in the third, ecological, phase.

An obvious, but difficult, part of any ecological investigation is to define the environment. This is especially true of work with soil, which so far as microbes are concerned, is a vast collection of micro-environments. However, there are 2 types of habitat which can be studied profitably; one is the plant root surface which will be dealt with in detail in a later paper (Brown, this Symposium), the other is the aggregate or crumb.

Many of the microbiological and chemical methods that have been used by soil workers are the same as, or similar to, those used by microbiologists in other disciplines and need no further comment here. However, some methods have been developed specifically for the microbiological analysis of soil; these need especial consideration.

2. The Estimation of Soil Microbial Populations

In any ecological study one must know not only which organisms are present but how many of them there are. The population of viable micro-organisms may be estimated in 2 ways, by counting colonies developing on or in agar medium inoculated with a dilute soil suspension, or by Most Probable Number (MPN) techniques; both of these general methods are well-known to microbiologists. MPN methods are rather inaccurate and require considerable replication of cultures if a reliable estimate is to be obtained; they are best reserved for counting particular groups of organisms, using suitable selective media.

Plate counting is often the only practicable method for estimating a soil microbial population. Though the method is usually accurate with a pure culture suspended in a clear diluent, it suffers from severe defects when used with a mixed population. For example, it is impossible to find one medium or one set of cultural conditions suitable for the growth of all micro-organisms; moreover, many cells are aggregated and do not disperse easily. These difficulties are well known but there is one defect of the method which is seldom emphasized: applied to soil, plate counting gives information about the soil suspension but tells us nothing about the arrangement of the organisms within the soil itself. The situation of the soil worker is basically different from that of the worker studying the occurrence of micro-organisms in, say, water. The latter needs to enquire whether certain organisms are present and in what quantity. So long as suitable medium and incubation conditions are available a complete answer will be obtained. In this instance the water is merely the vehicle for the organism and its nature is not determined by the organisms present. In contrast, organisms in soil are essential components of the system and their activities do much to create it. Soil is heterogeneous (Skinner, 1970) but the soil suspension on which the population estimate is based is, for statistical reasons, made as homogeneous as possible. Hence, even if accurate estimates could be made of all the organisms present one would still not be much wiser about the soil from which the suspension was made. Essential heterogeneity is lost.

The counting of micro-organisms in soil suspensions by direct microscopic observation is an alternative method to plate counting though it does, of course, estimate the total, instead of the viable, population. The particular method to be discussed is the Jones-Mollison direct counting technique, developed at Rothamsted in the late 1940s (Jones & Mollison, 1948). In this method a drop of soil suspension in molten agar is placed on a haemocytometer slide (0.1 mm deep), confined under a cover slip and allowed to set. The soil-agar film is then floated off to a microscopic slide and allowed to dry in air at room temperature, a process which causes no lateral shrinkage of the film. The dry film is then stained with acetic-aniline blue, washed, dehydrated quickly in alcohol and mounted in Euparal. When the mountant is hard the slides are examined by oil-immersion objective using fairly bright illumination and the stained cells in random fields are counted.

This method overcomes the selectivity inherent in plate counting and reveals the presence of much larger microbial populations in soils than are indicated by plate counts on the same samples. Many factors contribute to this discrepancy such as the direct counting of dead cells and cells that do not grow readily on the usual plate-counting media (Skinner, Jones & Mollison, 1952). Moreover, direct counting can give an estimate of total cells or a smaller estimate based on single cells plus cell clumps, that is, total colony-forming units. It is doubtless more reasonable to compare the latter estimate with plate count estimates.

Direct counting also yields interesting information about the distribution of types and sizes of cells and colonies in soil suspensions though it still tells us little about the spatial distribution of micro-organisms in relation to the soil particles.

3. The Distribution of Microbial Cell Sizes in Soil

The Jones-Mollison direct counting method has recently been used at Rothamsted (Powlson, 1972; Jenkinson, Powlson & Wedderburn, 1975) to obtain an estimate of microbial biovolume in soils. To do this these workers had to count the cells and also to estimate their volumes by using suitable graticules in the microscope eyepieces. This work was done as part of a larger project to test a hypothesis of Jenkinson (1966) who proposed that the size of the flush of organic matter decompositions following fumigation of soils with chloroform (and some other treatments) is proportional to the size of the biomass.

In order to disrupt clumps of cells and to release cells from occluding soil particles so that the most accurate estimates of numbers and sizes of cells could be made, soil suspensions were dispersed ultrasonically before admixture with the molten agar. This treatment gave higher total counts of soil micro-organisms, often double those given by the standard Jones-Mollison method in which the soil is ground only gently with water to prevent the disruption of small colonies.

For example, for one Rothamsted soil (Broadbalk, Plot 2) the standard technique gave a count of 3700×10^6 cells/g (1.99 mm^3/g) whereas, after ultrasonic treatment, the count increased to 5500×10^6/g with a biovolume of 4.96 mm^3/g. Spherical organisms (bacteria, spores of actinomycetes and fungi) were grouped into 13 diameter classes and cylindrical organisms (fragments of fungal and actinomycete hyphae) were grouped into 7 diameter classes: the frequency of occurrence of organisms of different size class was determined.

In the same soil c. 99% of the organisms (in the first 5 classes) were spherical and smaller than 1.6 μm diam. but they accounted for only 14% of the biovolume. Clearly one should not base estimates of biovolume on the size and number of typical bacterial cells. For spherical organisms there was an inverse relationship between the size of cell and numbers of cells; the smaller the cell, the more cells of that size. Moreover, if the spherical organisms are placed in volume classes divided geometrically (e.g. 0.1–1.0; 1.0–10.0 μm^3, etc.) each class accounts for the same amount of biovolume i.e. the total biovolume of spherical soil organisms with volumes between 1 and 10 μm^3 is the same as the total biovolume of the organisms with volumes between 10 and 100 μm^3. The results for 3 soils are given in Fig. 1.

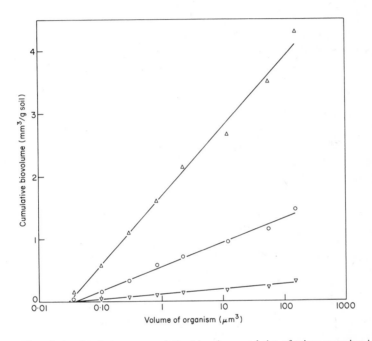

Fig. 1. The relationship between cumulative biovolume and size of micro-organism in soil. A point on a curve represents the total biovolume of all micro-organisms up to and including those of biovolume corresponding to that point. △, old grassland; ○, soil carrying continuous wheat crops (farmyard manure given anually); ▽, soil carrying continuous wheat crops (unmanured).

The total numbers of organisms and total biovolumes vary considerably between different soils, but despite this and differences in soil type and management, the distribution of numbers and biovolumes between the different classes is remarkably similar in all the soils tested. This is a valuable result because it indicates that these different soils are behaving in essentially the same manner in so far as their microbial populations are concerned. There is more than a hint here of a remarkable orderliness in soils which usually appear to be mixtures so complex as to defy description.

The significance of this relationship is not known but it would seem that more sites in the soil are able to support a small organism than a large one. Another way of looking at this is to say that, at any point in space and time, sufficient nutrients are available for just so much growth and that once this has occurred, time must elapse for fresh nutrients to diffuse to the depleted zone and, possibly, for metabolites to diffuse away from it.

Cylindrical organisms, including fragments of hyphae, do not obey this mathematically simple and satisfying relationship. The mode of growth of mycelial organisms is different; they ramify through the soil and exploit nutrients that they encounter. Moreover, the volume of a hyphal fragment, unlike that of a spherical organism, is not a characteristic of the species, but may merely reflect the number of times that the hypha was broken during preparation of the soil suspension.

An important matter to consider is whether this method gives a reasonably accurate measure of the biovolume or biomass. Calculations of biomass carbon derived from soil respiration measurements do, in fact, agree closely with estimates from biovolume data. Thus, 2 independent techniques reinforce each other and, if the only consideration is to measure biomass, there is good reason to use the faster and easier respirometer technique.

4. The Distribution of Microbial Colony Sizes in Soil Suspensions

Ultrasonic treatment of soil suspension before preparing the agar films disrupts small colonies and cell aggregates and releases cells from soil particles which can obscure them: this procedure is necessary for biovolume determinations or for obtaining maximum counts. However, by adhering to the original Jones-Mollison procedure which specifies only gentle grinding of the soil with water, one can obtain information on the distribution of micro-colonies of different sizes in the suspension.

In the resulting soil-agar films bacteria are distributed in colonies or clumps in a precise way according to a logarithmic series. This series is the expansion of the term $-\log_e (1 - x)$:

$$-\log_e (1 - x) = x + x^2/2 + x^3/3 + \ldots \ldots x^n/n$$

In terms of probability the series runs:

$$1 \text{ (i.e. certainty)} = \alpha x + \alpha x^2/2 + \alpha x^3/3 + \ldots \ldots \alpha x^n/n$$

where α is a constant equal to

$$1/-\log_e (1 - x).$$

The probability of finding colonies with 1, 2, 3 or n cells is given by the terms x, $x^2/2$, $x^3/3$ or x^n/n, respectively. In this expression x is less than unity and, in practice, the values found for soil films by Quenouille (in Jones & Mollison, 1948) lay between 0.7 and 0.8. The frequency of occurrence of a colony of a particular size is given by multiplying the appropriate probability term by the total number of colonies. The hundredth term of this series (representing a colony with 100 cells) is very small indeed and the thousandth term unimaginably small (Fig. 2).

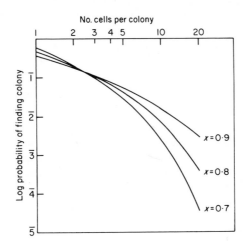

Fig. 2. Probable distribution of micro-colonies containing different numbers of cells in Jones-Mollison soil-agar films, according to the expansion of the term $-\log_e (1-x)$. Distributions are given for 3 values of x.

Thus there is little chance of finding a colony with even as few as 1000 cells. But such a colony is minute and if one were to obtain such colonies on an agar medium one would assume extremely unfavourable growth conditions for the organisms.

The logarithmic expression describes remarkably accurately the distribution of cells in the soil films but we must enquire whether the numerous single cells and small clumps actually exist in undisturbed soil or whether they are artefacts caused during preparation of the films. Clearly, many cell clumps in soil-agar films are colonies as may be inferred from their regular shape and because the cells composing them are often embedded in a common capsule. Such encapsulated masses would not be expected to break up when subjected to the mild grinding and mixing procedure needed to prepare the agar films. Other clumps are also obviously colonial, for example, short chains of bacteria or actinomycete spores which would be expected to break up readily. Some breaking of colonies must occur, but there is little evidence for the existence in soils of comparatively large colonies that separate into single cells as soon as the soil is stirred with water. Disruption of such large colonies would be expected to cause plate counts to approach more closely to the very large numbers characteristic of direct counts. One could, of course, postulate the existence of many large colonies of organisms that do not grow on the media generally employed for plate counts. This is unlikely, and there is no evidence for it.

There is other evidence that many organisms do not develop extensively in soils. Actinomycetes, for example, almost certainly exist frequently as spores. Fungus mycelium is found in direct count films but it normally occurs in short lengths and frequently stains faintly indicating that it is moribund.

On the whole, the weight of evidence is against the occurrence of large bacterial colonies or intensive development of mycelial organisms, and it is safe to assume that the logarithmic distribution characteristic of the direct count films represents with fair accuracy the situation in natural soils. Thus, there is no reasonable doubt that the growth of a single microbial cell into a colony is subject to severe limitation. It is nevertheless true that the chemical changes in soil are considerable and may be rapid. There are, after all, very many micro-organisms present and it would be remarkable if appreciable changes did not result from their activities. We are led therefore to the conclusion that the large chemical transformations in soil are resultants of the changes caused by vast numbers of individual cells of many species each of which is growing under conditions where colonial development is restricted. Undoubtedly, this restriction is due to a variety of causes such as shortage of nutrients available to any one cell, the presence of growth inhibitors and frequently generally adverse conditions of temperature and moisture.

It is interesting to compare the distribution of colony sizes in direct count soil films with those in agar pour-plates inoculated with soil dilutions. The result of one such investigation is given in Fig. 3. In this experiment, plates were prepared, each with 1 ml of a 10^{-4} soil suspension and 20 ml of an artificial soil solution agar (Skinner, Jones & Mollison, 1952) with or without glucose. After a few days the plates were flooded with acetic aniline blue stain for several hours, then washed to remove excess stain from the agar. Replicate discs, 1 cm diam, were removed from each plate and stored wet until they could be examined microscopically and the micro-colonies measured.

On these plates where colonies were deliberately overcrowded in a poor medium, conditions were unfavourable for growth. However, even the smallest colonies detected were very large compared with those found typically in direct count films, and the frequency distribution of the colony sizes clearly did not follow a negative logarithmic series. These findings emphasize the great constraints that are placed on micro-organisms in soil.

Thus we see how this direct counting technique, devised originally to give more accurate estimates of the soil microbial population, has extended our knowledge of micro-organisms in soil suspension.

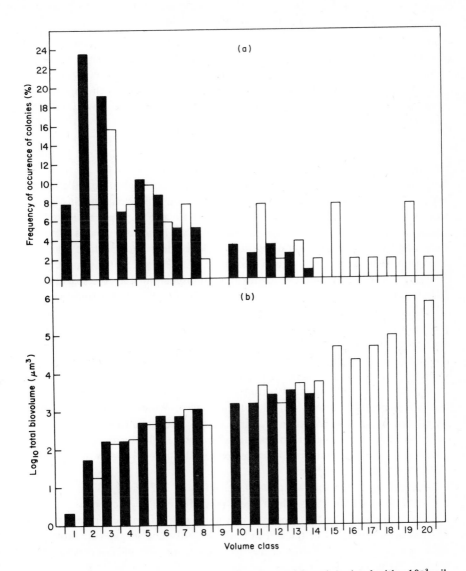

Fig. 3. The frequency distribution of colonies in pour-plates inoculated with a 10^{-3} soil suspension (1 ml of suspension + 20 ml of medium). ■, mineral salts agar without glucose; □, mineral salts agar with 10 g of glucose/l. (a) Frequency distribution of colony sizes with volume class. Class number is determined by diam of colony: thus, class 1 (up to 10 μm); class 2 (11–20); up to class 12: class 13 (121–150); class 14 (151–200); class 15 (201–250); class 16 (251–300); class 17 (301–400); class 18 (401–500); class 19 (501–700); class 20 (701–1000). (b) Distribution of total biovolume in each class shown in (a).

5. Microbial Ecology of the Soil Aggregate

Soil in not homogeneous with respect to its constituents or their arrangement: the ultimate particles of mineral and organic matter adhere to form aggregates of differing size and stability, the elements of soil structure. A top-soil with good structure consists mainly of aggregates separated by fine channels through which atmospheric oxygen diffuses to plant roots, and water percolates downwards for use by plants or to drainage.

(a) *Soil aggregates and aeration*

It might be thought that the small channels in soil, even when not filled with water, would offer considerable resistance to gaseous diffusion, but this is not so. Investigations on the changes in soil atmosphere at 6 in. depth by Russell & Appleyard (1915) at Rothamsted revealed that air in the soil spaces differed only slightly from atmospheric air in oxygen content though there was, on average, *c.* 10 times as much carbon dioxide (*c.* 0.25% in soil air as against 0.03% in the atmosphere). In addition to this free atmosphere of the pore spaces they found evidence of a second atmosphere, closely associated with soil colloids, which contained, typically, no oxygen and much CO_2, the remainder being nitrogen; probably the internal atmosphere of the soil aggregate or crumb. This atmosphere in close association with soil colloids indicated that the soil micro-organisms were using up oxygen faster than it could diffuse to regions of microbial activity remote from the larger soil spaces.

Figure 4 indicates the rapid diffusion of air through the larger channels so that the aggregates are well-aerated at their surfaces. Within the crumb, which is permeated mostly with extremely fine pores, gaseous diffusion is severely restricted. Even greater restriction occurs when a water film surrounding roots or microbial cells is encountered because oxygen diffuses through water 10 000 times slower than through gas.

Such an aggregated soil, when dry, takes up water in 2 stages; at first, water is absorbed by the aggregates which exert powerful capillary action because of their very fine pore size. This process continues with little effect on the gaseous diffusion characteristics of the soil as a whole until the aggregates are saturated when water begins to fill the spaces between them, and diffusion is quickly restricted (Currie, 1961).

The concept of a partially anaerobic aggregate in an aerobic environment is not new. At Rothamsted during World War I it was found that on cropped land which had received much organic manure there was a loss of nitrogen that could not be accounted for in the crop or as nitrate in the drainage water. Loss of nitrogen gas, resulting from denitrification, was suspected but could not be demonstrated by the analytical methods of the time.

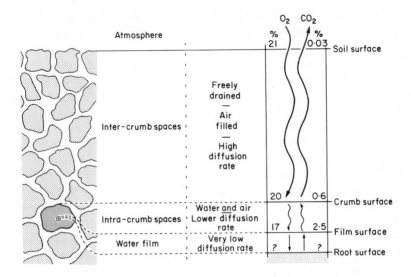

Fig. 4. Oxygen and carbon dioxide diffusion through the soil.

Concurrent studies on manure heaps (Russell & Richards, 1917) showed that nitrate was formed on the aerobic outside of a heap but that any nitrate washed into it would be reduced by facultative anaerobes, either to nitrate or gaseous nitrogen. Russell (1915) considered that the work on manure heaps could explain the loss of nitrogen from soil. He postulated that the soil aggregate behaved like a miniature manure heap in that nitrate could be formed at its aerobic surface but that the interior would tend to be anaerobic and would favour denitrification whenever nitrate diffused to it. Of course, the chance that an aggregate will become anaerobic depends on its size and constitution, as well as on the temperature and the availability of water. The diffusion characteristics of dry and wet aggregates have been the subject of much investigation and mathematical treatment in recent years (Currie, 1965; Greenwood, 1968).

The reducing conditions to be expected within a soil aggregate in which denitrification is taking place are indicated by the results of work with waterlogged soils. According to Bell (1969) a waterlogged soil usually remains at a fairly high redox potential ($c.$ + 200 mV) for a while, then falls to $c.$ − 200 mV when methane begins to appear in quantity in the fermentation gases. When the soil contains nitrate the E_h remains at $c.$ + 200 mV until all nitrate has been denitrified. Thus it seems unlikely that the interiors of soil aggregates will become highly reducing when nitrate is present even though they may be anoxic.

Although lack of oxygen leads to the undesirable result of denitrification it may have some beneficial results as well. Recently, a detailed study was made of non-symbiotic nitrogen fixation occurring in soils of Broadbalk field at Rothamsted and at neighbouring sites including Broadbalk Wilderness, an area originally sown to wheat but allowed to revert to woodland at the beginning of this century; some parts of it are cut annually. Soil samples were taken from the latter areas when under thick vegetation cover as well as from the arable experimental plots. Nitrogen-fixing ability was estimated from nitrogenase activity of the soil using the acetylene reduction technique.

Considerable nitrogenase activity was found in soil cores under vegetation cover. There is evidence that non-symbiotic nitrogen-fixing organisms, especially facultative anaerobes of the *Klebsiella* type, were active in the rhizospheres of the plants, especially *Stachys sylvatica*, *Mercurialis perennis* and *Heracleum sphondylium*. Nitrogenase activity was highly and positively correlated with soil water content. It is postulated that as soil moisture increases, the level of anaerobiosis in the soil aggregates and rhizospheres increases and stimulates nitrogenase activity; there is probably no direct effect of the water on the nitrogenase. The arable areas of Broadbalk showed little bacterial nitrogen fixation probably because, under cultivation, the soil is rarely wet enough to support appreciable fixation in the rhizosphere (Day *et al.*, 1975).

It should be noted that the water holding capacity of Broadbalk soil has a mean value of *c.* 25% (at 50 cm suction) so samples containing more than this amount of water will be waterlogged to a greater or lesser extent. However, nitrogenase activity increased sharply between the values of 0 and 25% water content suggesting increasing activity as the crumbs became increasingly deficient in oxygen.

(b) *Distribution of micro-organisms within aggregates*

As conditions of aeration and moisture at the surface of a crumb are likely to differ from those inside it, we must expect differences in the microflora of these 2 zones. Some attempts have been made to fractionate aggregates in ways favourable to subsequent microbiological analysis but the methods have not been used extensively.

As Holt & Timmons (1967) showed, it is not difficult to separate the outer layers of crumbs from the inner parts, for chemical analysis. They did this by freezing the crumbs, then partially thawing them so that the outer parts could be washed away from the still frozen centres.

A somewhat analogous method of microbiological analysis was developed by Tyagny-Ryadno (1968). In this, the dry aggregates to be fractionated are treated with collodion so that each is covered with a thin layer of this material. The treated aggregates are then shaken in water for *c.* 5 min to break them up; the collodion films carrying the outer layers of the aggregates float to the surface and are removed. Appropriate microbiological analyses are performed on the remaining suspension which contains organisms derived from the interiors of the aggregates. Similar tests are made on suspensions of whole aggregates and the microbial content of the outer layers determined by difference.

Results (Tyagny-Ryadno, 1962, 1968) showed that fungi and ammonifiers were more numerous in the outer, than in the inner, layers of aggregates, though aerobes were also present in the interiors. Clearly, this method shows up differences between different parts of aggregates. However, collodion has to be dissolved in a mixture of ethanol and ether, either of which could have adverse effects on the micro-organisms within the crumbs during the coating procedure.

The method of Hattori, described by Nishio, Hattori & Furusaka (1968), relies on the fact that water-stable aggregates can be washed gently to remove organisms from their outer layers. The interiors can then be suspended in diluent by more vigorous treatment. A weighed quantity (e.g. 2 g) of aggregates is added to 100 ml of sterile water in a 250 ml flask, shaken gently by hand for 1 min and allowed to settle for 1 min. One ml of the supernatant liquid containing organisms from the outer layers of the aggregates is removed for analysis and the remainder discarded by decanting. This washing procedure is repeated twice, though most of the superficial micro-organisms are released by the first wash. Finally the aggregates are mixed with 50 ml of water and sonicated for 3 min. The sonicate is diluted and used for microbiological analysis of the inner parts of the aggregates.

Nishio et al. (1968) inoculated sterilized aggregates (0.83–2.0 mm diam) of a garden soil with a diluted broth culture of *Escherichia coli* and used the Hattori method to follow growth and distribution of the organism. Growth of the introduced cells followed similar patterns in the 2 fractions which is what one would expect with a facultative anaerobe. However, there were differences in rates of growth in the different fractions, presumably reflecting differences in the physical conditions between the outer and inner parts of the aggregates. Thus cells in the outer fraction grew faster than those inside the aggregate when the soil had a water content of 40% of the water holding capacity (WHC), but grew slower than those inside at the higher water content of 60% WHC.

Nishio (1970) also studied the distribution of nitrifying bacteria in soil aggregates by the same method. When aggregates were percolated with solutions containing ammonium or nitrite ions, nitrifying bacteria proliferated more in the outer, than in the inner parts. This agrees with what is known about the aerobic nature of the organisms and supports the early ideas of Russell (1915). There is no doubt that this method has given interesting results in the hands of Japanese workers. However, it is difficult to see how it could be used with any but the most water-stable of aggregates.

This is a convenient point to refer to a method of dry fractionation that I am now investigating. The method depends on the fact that the outer layers of aggregates are mutually abraded when they are shaken together. The apparatus consists of a pair of 5 ml Bijou screw-capped bottles linked together by a Hemming filter assembly without the filter pad. Crumbs are placed in the lower container and the whole shaken vertically in a wrist-action shaker. After the required period the assembly is inverted and shaken gently to allow the fine particles to pass downward into the receiver. The latter is then replaced by another empty, weighed, sterile bottle and the shaking resumed as required. The fractions are weighed, suspended in diluent and analysed. The outermost material is quickly removed, some 10% of the original weight in the first 3 min under the experimental conditions used; as the crumbs become smoother and rounder it becomes increasingly difficult to detach material from the crumbs. Of course, only fairly dry crumbs can be used in this method.

In view of the fact that the physical properties of soil aggregates change according to moisture content, it will doubtless be necessary to develop and use several complementary methods for investigating the distribution and activity of the aggregate microflora.

(c) *Formation of soil aggregates*

Micro-organisms are also involved in the formation, stabilization and degradation of soil aggregates, and an extensive literature exists as shown in the review by Harris, Chesters & Allen (1966).

Soil particles can be bound together mechanically by fungal or actinomycete hyphae (Swaby, 1949), but the resulting aggregates, though fairly stable, last only until the hyphae are decomposed by other micro-organisms.

There seems little doubt that microbial polysaccharides play an important part in linking soil particles. These substances are produced by many micro-organisms (Hepper, 1975), and soil organic matter may contain up to 25% of them (Martin, 1971). In a free state, polysaccharides are easily degraded by other micro-organisms but they seem to be protected from such degradation once they are bound firmly within an aggregate or incorporated in a clay lattice. Humified organic residues, though highly resistant to microbial decomposition, seem to exert little binding action on soil particles. Many laboratory investigations have been made with a view to determining the role played by micro-organisms in the formation of stable soil aggregates. However, aggregates made in the laboratory usually do not possess the stability characteristic of those formed naturally.

Allison (1968) has distinguished between factors that lead to aggregate formation and those that confer stability to the action of water. He comments on the well-known fact that stable aggregates tend to be formed under grass and suggests that this comes about by the combined action of plant and microbe. The numerous fine roots ramify and divide up the intervening soil into small blocks, exert pressures, and by removing water cause local drying. At the same time root exudates and sloughed root tissue support growth of micro-organisms and promote the formation of polysaccharides just where they are needed to stabilize the newly-formed aggregates. Such aggregates are to some degree stable to mechanical forces and very resistant to water.

The methods so far used in microbiological work on the formation and decay of aggregates have been simple ones involving, for the most part, inoculation of soil with living organisms or treatment with microbial products, and assessing the amount of water-stable aggregates by determining their stability when sieved under water or when subjected to falling water drops.

6. Conclusions

The soil microbiologist is faced with a daunting situation. He has to deal with a system which is heterogeneous physically and chemically, vast in extent, and highly diverse in origin, mode of formation and in the microbial populations that it contains. Typically, the microbial population is large and complex but not necessarily very active. The evidence indicates that growth of micro-organisms in soil is subject to severe restriction, local scarcity of nutrients in micro-environments being probably the major cause. One should not regard soil as a medium for microbial growth but rather as an environment that protects and favours survival of a varied population with a wide range of biochemical ability.

The physical nature of soil precludes the easy diffusion of soluble substrates and prevents rapid distribution of the many insoluble components of plant and animal residues. Agencies likely to bring together organic matter and the micro-organisms are of obvious importance and in this connection we must not underestimate the mixing and comminuting activities of the soil fauna.

The soil microbiologist needs to make use of all basic methods known to microbiologists generally, but in addition, he must try to understand the complex physico-chemical nature of the soil and develop special methods that this particular environment makes necessary.

In this present, ecological, phase of soil microbiology I have tried to show how important it is to define the free soil environment and to relate microbial activity to it. Soon, I hope a phase of synthesis will begin, characterized by a greater readiness to study the vast amount of information already available and to co-operate with soil scientists in other disciplines. It is our function to understand the principles on which soil microbial populations work, and how soil processes are affected by soil type, climate and husbandry, in order to predict behaviour of soils and thus to advise on the ways that they should be used to preserve them in as productive a state as possible.

7. Summary

In this consideration of methods used by soil microbiologists emphasis is placed on those methods that have been developed especially for work with soil, and on those useful for ecological studies. The use of these methods is exemplified by an analysis of the factors influencing the distribution of cells and micro-colonies in soils and soil suspensions. The microbial ecology of soil aggregates is discussed in relation to aeration, denitrification and soil structure.

8. Acknowledgements

I am indebted to Dr D. S. Jenkinson and to Dr J. A. Currie for permission to reproduce Figs 1 and 4, respectively. I also thank them both for many helpful discussions on soil chemistry and physics.

9. References

ALLISON, F. E. (1968). Soil aggregation – some facts and fallacies as seen by a microbiologist. *Soil Sci.* **106**, 136.

BELL, R. G. (1969). Studies on the decomposition of organic matter in flooded soil. *Soil Biol. Biochem.* **1**, 105.

CROSS, T. & GOODFELLOW, M. (1973). Taxonomy and classification of the actinomycetes. In *Actinomycetales: Characteristics and Practical Importance,* Society for Applied Bacteriology Symposium Series No. 2. Eds G. Sykes & F. A. Skinner. London & New York: Academic Press.

CURRIE, J. A. (1961). Gaseous diffusion in the aeration of aggregated soils. *Soil Sci.* **92**, 40.

CURRIE, J. A. (1965). Diffusion within soil microstructure. A structural parameter for soils. *J. Soil Sci.* **16**, 279.

DAY, J. M., HARRIS, D., DART, P. J. & VAN BERKUM, P. (1975). The Broadbalk experiment. An investigation of nitrogen gains from non-symbiotic fixation. In *IBP Synthesis Volume on Nitrogen Fixation.* Cambridge: Cambridge University Press.

GREENWOOD, D. J. (1968). Measurement of microbial metabolism in soil. In *The Ecology of Soil Bacteria,* Symposium, Liverpool, 1965. Eds T. R. G. Gray & D. Parkinson. Liverpool: University Press.

HARRIS, R. F., CHESTERS, G. & ALLEN, O. N. (1966). Dynamics of soil aggregation. *Adv. Agron.* **18**, 107.

HATTORI, T. (1966). On numerical variation of Gram-negative bacteria in soil aggregates. *J. Sci. Soil Manure, Japan* **37**, 298.

HEPPER, C. M. (1975). Extracellular polysaccharides of soil bacteria. In *Soil Microbiology.* Ed. N. Walker. London & Boston: Butterworths.

HOLT, R. F. & TIMMONS, D. R. (1967). A method for investigating the chemical heterogeneity of soil material within natural soil aggregates. *Proc. Soil Sci. Soc. Am.* **31**, 704.

JENKINSON, D. S. (1966). Studies on the decomposition of plant material in soil. II. Partial sterilization of soil and the soil biomass. *J. Soil Sci.* **17**, 280.

JENKINSON, D. S., POWLSON, D. S. & WEDDERBURN, R. W. M. (1975). The effects of biocidal treatments on metabolism in soil. III. The relationship between soil biovolume, measured by optical microscopy, and the flush of decomposition caused by fumigation. *Soil Biol. Biochem.* (In Press).

JONES, P. C. T. & MOLLISON, J. E. (1948). A technique for the quantitative estimation of soil micro-organisms. *J. gen. Microbiol.* **2**, 54.

MARTIN, J. P. (1971). Decomposition and binding action of polysaccharides in soil. *Soil Biol. Biochem.* **3**, 33.

NISHIO, M. (1970). The distribution of nitrifying bacteria in soil aggregates. *Soil Sci. Plant Nutr.* **16**, 24.

NISHIO, M., HATTORI, T. & FURUSAKA, C. (1968). The growth of bacteria in sterilized soil aggregates. *Rep. Inst. Agr. Res. Tohoku Univ.* **19**, 37.

POWLSON, D. S. (1972). The effects of fumigants on soil respiration and mineralization of nitrogen. Ph.D. thesis, University of Reading.

ROSSWALL, T. (Ed.) (1973). *Modern Methods in the Study of Microbial Ecology.* Bulletin No. 17, Ecological Research Committee. Stockholm: Swedish Natural Science Research Council.

RUSSELL, E. J. (1915). Director's Report. *Rep. Rothamsted Exp. Stn. for 1914.*

RUSSELL, E. J. & APPLEYARD, A. (1915). The atmosphere of the soil: its composition and causes of variation. *J. agric. Sci., Camb.* **7**, 1.

RUSSELL, E. J. & RICHARDS, E. H. (1917). The changes taking place during the storage of farmyard manure. *J. agric. Sci., Camb.* **8**, 495.

SCHATZ, A., BUGIE, E. & WAKSMAN, S. A. (1944). Streptomycin, a substance exhibiting antibiotic activity against gram-positive and gram-negative bacteria. *Proc. Soc. exp. Biol. Med.* **55**, 66.

SKINNER, F. A. (1970). Microbial heterogeneity of soil. In *Soil Heterogeneity and Podzolization.* Ed. E. M. Bridges. *Report Welsh Soils Discussion Group* No. 11.

SKINNER, F. A., JONES, P. C. T. & MOLLISON, J. E. (1952). A comparison of a direct- and a plate-counting technique for the quantitative estimation of soil micro-organisms. *J. gen. Microbiol.* **6**, 261.

SWABY, R. J. (1949). The relationship between micro-organisms and soil aggregation. *J. gen. Microbiol.* **3**, 236.

TYAGNY-RYADNO, M. G. (1962). The microflora of soil aggregates and plant nutrition. (Russian, English summary). *Izvest. Akad. Nauk USSR. Ser. Biol.* **2**, 242.

TYAGNY-RYADNO, M. G. (1968). Distribution of micro-organisms and nutrients in soil aggregates. (Ukrainian, Russian summary) *Visn. sil. – hospod. Nauky.* **11**, 46.

WAKSMAN, S. A. & WOODRUFF, H. B. (1940). The soil as a source of micro-organisms antagonistic to disease-producing bacteria. *J. Bact.* **40**, 581.

Microbial Manipulation and Plant Performance

MARGARET E. BROWN

*Soil Microbiology Department, Rothamsted Experimental Station,
Harpenden, Hertfordshire, England*

CONTENTS

1. Introduction

MOST SOIL MICROBIOLOGISTS are interested in micro-organisms that improve soil fertility, and by studying those found in the root region hope to gain a better understanding of how this microflora affects plants. The disciplines of plant pathology and soil microbiology are combined when growth of root pathogens in relation to the rhizosphere flora is studied and the potential of biological control of plant diseases is explored.

Before biological control of rhizosphere populations can be achieved, at least 2 basic phenomena must be clarified. First, knowledge is needed of organisms colonizing plant roots at any given time under a given set of environmental conditions and of the mechanisms governing that colonization. Secondly, knowledge of the function of rhizosphere populations is required. In 30 years of intensive study of rhizosphere ecology several main lines have been followed. That receiving most attention has been the effect of the plant on the micro-organisms, especially in inducing qualitative and quantitative changes in the composition of the population, and a direct consequence of this has been the study of root exudates. Recently the effects of the microbes on the plants and inter-relationships between groups of organisms have been studied, followed by attempts to manipulate the rhizosphere microflora, principally by altering root exudates or changing the soil environment.

2. The Rhizosphere

The soil microbial population lives in an unstable equilibrium where at any one time all individuals are in balance with each other but where changes in the environment lead to changes in the equilibrium. Plant roots provide such an environmental change and with their rich supply of nutrients immediately stimulate micro-organisms especially at the root surface where numbers often reach 10^{10}/g of root surface soil. This population decreases with increasing distance from the root, the extent of the gradient depending principally on plant species. Bacteria, fungi, actinomycetes, free-living nematodes are all stimulated selectively (Rovira, 1965a).

Micro-organisms begin to colonize the root within a few hours of germination, most of the population coming from the soil microflora rather than from the seed coat (Rovira, 1956). Bacteria develop in the zone of root elongation, initially as widely separated clusters of cells but as roots age, colonies increase in size and may eventually form a mantle over the root (Rovira & Campbell, 1974). Studies on spatial distribution of micro-organisms on roots of field-grown wheat showed that seminal roots had few colonies occurring mainly at junctions of cortical cells; young nodal roots were coated with a uniform layer of soil containing few bacteria, and old nodal roots had large aggregates of bacteria both inside and out (Foster & Rovira, 1973).

Fungi do not colonize roots until at least 24 h after germination; dormant spores in the soil are stimulated and probably successive lateral root colonization is more important than longitudinal growth of mycelium down the roots starting from an initial point. Mycelial activity increases as plants age; in middle life large areas of root are still free of fungi but with maturity all the spaces and internal tissues become colonized (Parkinson, 1967). Although fungi are isolated less frequently than bacteria their importance as active members of the root surface population may be just as great, because a few units of large hyphae are equivalent to thousands of bacteria.

3. Root Exudates

As roots move through the soil the root cap cells provide nutrients near the tip and elongating zone. As plants age, root hairs and cortical cells die followed by the whole root and all this material feeds micro-organisms. The cell debris is probably less selective than root exudates because older roots support a large and varied population of species, not all of which are found in the select community associated with young healthy roots (Rovira, 1965b). Although intact roots exude relatively small amounts of organic material they produce enough to support enormous microbial populations; the amount determines the size and the type of material affects the nature of the microflora (Rovira, 1969).

Bowen & Rovira (1973) found that roots of 2- and 8-week-old wheat grown in the field supported 1.5×10^6 bacteria/mg of dry root. Root exudates contain small amounts of a miscellany of compounds including sugars, amino acids, peptides, enzymes, vitamins, organic acids, nucleotides and in trace amounts, various substances with specific biological activities such as nematode cyst-hatching factor and fungal zoospore attractants (Rovira, 1969). Some of these may diffuse away from the root or be adsorbed by the soil, but calculations of expected microbial populations based on the energy content of these materials agree well with actual counts made by orthodox dilution plate techniques (Bowen & Rovira, 1973).

Different plants exude different amounts and types of substance (Vancura & Hanzlíková, 1972) and exudation from ony one plant will differ according to its nutrition (Street, 1969). Other environmental factors influence the pattern of exudation, for example, high light intensity and temperature (Rovira, 1959), temporary wilting (Katznelson, Rouatt & Payne, 1955) and root damage (Rovira, 1969) all favour exudation.

4. Effects of the Rhizosphere Microflora

(a) *Availability of nutrients*

Plant roots stimulate certain bacteria selectively, these are mostly Gram negative rods and pleomorphic forms belonging to dominant groups that either ferment carbohydrates, decompose cellulose, ammonify or denitrify (Katznelson, Lochhead & Timonin, 1948; Starkey, 1958; Rovira, 1965). These bacteria are more active physiologically as shown by manometric experiments, when increased rates of oxygen uptake are obtained with substrates such as sucrose, glucose, acetate, succinate or alanine (Zagallo & Katznelson, 1957). However, abundance of any one group does not necessarily mean greater activity; for example, although denitrifiers are stimulated there is no evidence that denitrification is generally more rapid in the rhizosphere, but it may be important in local areas of anaerobiosis leading to losses of fertilizer nitrogen supplied as nitrate. Equally, activity of ammonifying bacteria may lead to increased ammonia concentrations in the rhizosphere. Recent work by the author (unpublished) has shown that when wheat was actively growing, ammonium nitrogen was in excess of nitrate nitrogen in the rhizosphere, but not in soil some distance from the roots. When growing rapidly wheat makes great demands on the nitrate supply resulting in nitrate stress in the soil (Nair & Talibudeen, 1973). Such nitrogen supply conditions must affect the activity of the root microflora and Goring & Clark (1948) showed that there was less nitrogen in the rhizosphere than would have been formed as nitrate in unplanted soil. Nitrification was proceeding, but at the same time nitrogen was being immobilized by the micro-organisms.

Micro-organisms can affect nutrient uptake by roots by changing the availability of nutrients and modifying uptake processes. Minerals are required for microbial growth and when in short supply will be used rapidly by the mantle of micro-organisms outside the root, to the detriment of the plant. Such immobilization of zinc or manganese leads respectively to the deficiency diseases 'little leaf' of fruit trees and 'grey speck' of oats (Barber, 1968).

An interesting example of the effect of the rhizosphere flora on molybdenum contents of plants is described by Loutit and her co-workers (Loutit, Loutit & Brooks, 1967; Loutit, Malthus & Loutit, 1967; Loutit & Brooks, 1970). A dental survey in the Hawke's Bay area of New Zealand showed that dental caries differed in 2 districts, Napier and Hastings, and it was suggested that the caries was related to molybdenum deficiency in vegetables grown in Hastings soil. These vegetables had lower concentrations of aluminium, molybdenum and titanium and higher concentrations of barium, copper, manganese and strontium than vegetables grown in Napier soils. The soils of both areas were derived from a similar parent rock so it was thought that microbial activity in the rhizosphere had affected mineral uptake. Experiments with radish showed that numbers of micro-organisms in the rhizospheres of plants grown in the 2 soils were similar, but when plants were grown in nutrient solutions and inoculated with rhizosphere organisms from Hastings soil, they contained less molybdenum than if inoculated with organisms from Napier soil.

The effect of the rhizosphere microflora on uptake and translocation of phosphorus has attracted considerable interest. Sometimes a very high proportion of organisms are able to dissolve water-insoluble inorganic phosphates, such as hydroxyapatite, by secreting organic acids (Swaby & Sperber, 1959). Plant roots, and possibly micro-organisms, may also bring phosphate ions into solution through uptake of calcium ions from sparingly soluble calcium phosphates. The uptake of calcium causes a shift in the mass action equilibrium (Johnston & Olsen, 1972). Such a solution process would be favoured by close contact with the surfaces of mineral particles, which is achieved more readily by bacteria and fungi, because of their small size, than by plant roots. Fungi obtain more phosphate when their hyphae are in contact with particles of finely ground minerals than when contact is interrupted (Swaby & Sperber, 1959).

Organisms producing phosphatases are sometimes stimulated in the rhizosphere and release orthophosphate from both organic and inorganic phosphates (Greaves & Webley, 1965). Available evidence, however, suggests that organic phosphates, such as phytate, accumulate in soil and Greaves & Webley (1969) showed that soil inhibited hydrolysis of sodium phytate by micro-organisms that could attack it readily in sand. Equally, organisms able to mineralize organic phosphate in culture could not do so in soil, nor was there conclusive evidence that soil micro-organisms were directly involved in liberating inorganic phosphate.

The overall impression from the extensive literature is that microbial mineralization of organic phosphate occurs widely in soil, but how much of this phosphate is surplus to the bacterial requirements and hence available to plants is not known. In normal cultivated land it is estimated that bacteria alone take up 4–10 kg of P/hectare (Sauchelli, 1965). When fungal and actinomycete activity is also taken into account it seems that micro-organisms remove at least as much phosphorus as the crop. The phosphate immobilized within microbial cells is released when they die. When plant residues decompose, the microbial population needs more phosphate, and if the carbon-phosphorus ratios are high, micro-organisms may assimilate so much phosphate that crop yield is depressed, unless fertilizer is added. Decomposing organic matter has a critical concentration of phosphorus of c. 0.2%, above and below which there is net mineralization or immobilization, respectively (Alexander, 1961).

Some good evidence that phosphate solubilization benefits plants comes from experiments where adding straw to soil increased phosphate uptake both from added insoluble salts and from sparingly soluble native soil phosphate. The microbially degraded straw released 'humic' and 'fulvic' acids which formed stable complexes with iron and phosphate that were as available to plants as orthophosphate ions (Mishustin, 1972).

(b) *Activity of growth regulating substances*

The rhizosphere microflora affects plants by producing growth regulating substances. Swaby (1942) observed that plants grown with micro-organisms could be twice as tall as those grown aseptically and suggested that in association with organic matter micro-organisms frequently produced both stimulating and inhibiting substances. Subsequent work (Gruen, 1959) showed that many organisms formed indolyl-3-acetic acid (IAA) in culture media, especially in the presence of tryptophan. Recently Brown (1972) found that young wheat roots supported many bacteria able to produce IAA and this population decreased as plants aged (Table 1). Thus these bacteria were abundant when tryptophan was likely to be one component of root exudates. The amounts of auxin produced by the cultures ranged from 0.02–0.3 μg/ml without tryptophan, and from 1–10 μg/ml when 100 mg/l of L-tryptophan was added to the medium. *Pseudomonas* spp. are particularly active producers and are also stimulated in the rhizosphere of wheat seedlings, but not of older plants.

There is now ample evidence that roots take up a variety of compounds including indoles, and the region of uptake is probably the zone of differentiation or where root hairs are most abundant, that is, the zone where the root microflora is also richest. Uptake of exogenous IAA increases the auxin content of the tissues (Scott, 1972) and root growth is partly controlled by auxin; decrease in growth rate may be an indirect effect due to auxin-induced production of ethylene, which is thought to be the immediate inhibitor. Usually, inhibited roots are thickened resulting from interaction with cytokinin. Very small doses of IAA may give growth without inhibition and sometimes brief exposure to auxin causes inhibition, followed by stimulation (Thimann, 1972).

Table 1

Percentage of isolates from wheat root region and soil producing
IAA with and without L-tryptophan in the medium

| | % isolates producing IAA | | | | | |
| | Without L-tryptophan | | | Producing more IAA with L-tryptophan | | |
Days	Rhizosphere	Rhizoplane	Soil	Rhizosphere	Rhizoplane	Soil
6	77	84	87	35*	54*	29
42	73	78	86	47*	7	29
82	89	88	87	4	0	28

* Denotes significant difference from soil population ($P = 0.05$).

Microbial production of small amounts of IAA in the rhizosphere is thus likely to alter plant root growth, and various monoculture tests with different organisms have given results indicative of IAA effects. Bowen & Rovira (1961) found total length of several plant species was nearly always significantly decreased by soil microorganisms, the concentration of secondary roots on primary roots was greater, and root hairs at the proximal part of the root were shorter and less abundant. Maize grown in unsterile nutrient solution was increased in length and weight because the bacteria produced IAA which was taken up by the plant (Libbert & Manteuffel, 1970). *Azotobacter chroococcum* causes a general enlargement of the root system with longer roots and more laterals (Brown, unpublished). Differences in effects on roots may be caused by different amounts of hormone in the cultures or by production of more than one growth regulator. Strzelczyk, Kampert & Krzysko (1971) had evidence of interactions of growth regulators in experiments with *Arthrobacter pascens* and *A. globiformis*. *Arthrobacter pascens* cultures contained substances that in the *Avena* test act as inhibitors, and as inhibitors of auxin- and gibberellin-induced plant growth. Lettuce assay plants were stimulated by these cultures. It has now been shown that *Arthrobacter* spp. can produce cytokinins (Blondeau, 1970).

In comparing combined effects of the regulators IAA and gibberellic acid (GA_3) with the effects of soil micro-organisms on wheat seedlings growing in perlite moistened with nutrient solution, Brown (1972) found marked similarities, suggesting that microbial production of regulators was implicated in the effects on the seedlings (Table 2).

Soil micro-organisms produce gibberellin-like substances in sufficient quantity to affect plants. The term gibberellin-like generally refers to incompletely characterized substances, or substances not containing the gibbane nucleus but which have biological activity in gibberellin bioassay systems. As little as 0.005 μg of authentic GA_3 applied to roots of seedling tomatoes at the cotyledon stage of development, a critical stage of differentiation of primordia, is enough to alter internode length, leaf area and time of flowering (Brown, Jackson & Burlingham, 1968) and many organisms produce at least this amount of gibberellin-like substance. Many bacteria produce a minimum of 3 substances showing activity in dwarf pea and lettuce bioassays and equivalent to 0.001–0.5 μg GA_3/ml of culture (Brown, 1972). Such bacteria are particularly abundant in rhizospheres of wheat at the stage of tillering or panicle emergence.

Table 2

*Comparison of effects on wheat seedling growth of treatment
with GA_3 + IAA, soil or water*

	Effect of treatment on seedling growth		
	Control (water)	GA_3 + IAA	Soil suspension
Leaf No.	2	2.8*	2.7*
Total length (mm)	373	470*	435*
Total weight (mg)	2.44	2.98*	2.97*
Root No.	6	6.8*	5.5*
Total length (mm)	1379	1174*	1163*
Total weight (mg)	2.12	1.69*	1.43*

* Denotes significant difference from control (P = 0.05).

Seedlings were grown in perlite (Peralite; British Gypsum, Mountfield, Robertsbridge, Sussex) moistened with mineral salt solution. GA_3 and IAA were supplied at the rates of 0.01 and 0.05 μg/seedling, respectively.

External application of gibberellin can replace environmental and genetic factors otherwise required for normal plant development. Primarily, gibberellins induce cell division but can also affect cell expansion, and their pronounced effect is to increase growth rate of dwarf stems with some increase in leaf area. By altering the pattern of cell division vegetative apices can be changed into floral apices, flowering can be induced in rosette plants and sex of flowers altered. Gibberellin treatment can also break seed dormancy leading to germination. These are some of the induced effects, so clearly microbial production of hormones with these properties will modify plant growth.

Several micro-organisms as well as *Arthrobacter* have now been shown to produce cytokinin-like substances, *Azotobacter chroococcum* and *A. paspali*

(Coppola *et al.,* 1971; Barea & Brown, 1974), *Agrobacterium tumefaciens* (Romanow, Chalvignac & Pochon, 1969), *Corynebacterium fasciens* (Helgeson & Leonard, 1966), *Rhizobium japonicum* (Phillips & Torrey, 1970), and cyto-kinins are present in RNA preparations from several other species (Skoog & Armstrong, 1970). Cytokinins are involved in all phases of plant development from cell division and enlargement to formation of flowers and fruits. They affect metabolism and biosynthesis of growth factors, influence appearance of organelles and flow of assimilates and nutrients through the plant, and also enhance resistance to ageing and to adverse environments.

Micro-organisms such as *Azotobacter* spp. produce IAA, gibberellin and cytokinin-like substances and it is not unreasonable to suppose that such metabolites produced in the root/soil environment may together influence plant development more than one substance alone.

5. Manipulation of the Microflora

Using the information gathered over the years on activity of the rhizosphere microflora, microbiologists are now attempting to manipulate the flora to one selected for properties that benefit plants either by direct effect or by controlling plant diseases. Three main ways are possible; (a) by adjusting environmental conditions, (b) by altering plant metabolism, (c) by introducing different organisms into the rhizosphere.

(a) *Adjustment of the environment*

Adjusting environmental conditions by appropriate crop rotations and soil management is one of the oldest and best known examples of biological control. Inoculum density of the pathogen is generally lowered, either by encouraging growth of antagonists, or growth of a more general microflora that competes with the pathogen for nutrients. Green manuring permits growth of rhizosphere micro-organisms in soil distant from roots, thus widening the zone of competition and antagonism around the plant.

Control of potato scab, caused by *Streptomyces scabies,* by green manuring is a good example of antagonism of the pathogen and how quality of the green manure can affect the amount of antibiotic produced. Soybean, but not barley, grown as a green manure crop in the autumn after potato harvest and turned in before spring planting of potatoes controls the disease. Both green manures encourage growth of *Bacillus subtilis,* but only soybean tissue is a suitable substrate for production of the antibiotic which probably inhibits the streptomycete (Weinhold & Bowman, 1968).

Adding, to soil, crop residues that lead to competition between pathogen and the microflora often gives very good disease control. Many root diseases are caused by fungi normally present as spores that require nutrients for germination and infection. Thus a pathogen requiring nitrogen can be suppressed by adding residues with high C : N ratios because competition renders nitrogen in short supply. Many fungal propagules are controlled by a fungistatic factor in soil preventing germination and mycelial growth even in favourable conditions; this factor is often overcome by nutrients, especially those found in the rhizosphere (Lockwood, 1964; Schroth & Hildebrand, 1964). The phenomenon of fungistasis is very important economically, for induction, maintenance and release from fungistasis are phases through which a pathogen has to pass to maintain its inoculum potential between susceptible hosts. Interference with this process can cause decline of the pathogen, for example, an amendment that overcomes fungistasis and stimulates pathogen growth in the absence of the host will cause death of the inoculum by starvation.

Changing the ecological balance of the soil, leading to substantial disease control, has been achieved by adding certain selective chemicals which in themselves are not toxic to the pathogens. Field applications of the monosodium salt of hexachlorophene at 280 g/ha to cotton plants and soil controlled several diseases caused by *Rhizoctonia, Pythium, Phytophthora* and *Fusarium* (Pinckard, 1970). Similarly, cucumber was protected from damping-off by soaking the seed in a 1 p/m solution of 6-azauracil. This concentration changed the balance of the rhizosphere population to one of more bacteria and fewer fungi; other concentrations were ineffective (Stankova-Opocenska & Dekker, 1970).

(b) *Changing the metabolism of the plant*

Quantitative and qualitative changes in root exudates follow foliar applications of nutrients, antibiotics and growth regulators, and as a consequence the microbial population changes. The actual mechanisms involved differ, for example, antibiotics are usually translocated through the plant and exuded unchanged, but nutrients such as urea are metabolized and exuded in a changed form (Vraný, Vancura & Mačura, 1962). Vraný (1965) found that applying urea increased the number of bacteria and decreased the fungi in the rhizosphere. Rhizosphere nutrition is also altered when different nutrients are applied to the roots. Brown (unpublished) found in axenic cultures of wheat that some strains of *Gaeumannomyces graminis* var. *tritici,* causing take-all disease, were less able to infect seedlings grown with ammonium than when grown with nitrate. Root exudates from ammonium grown plants support less fungal growth than those from nitrate grown plants. This is not an effect of pH but of the host nutrition.

(c) *Introduction of organisms to the rhizosphere*

Modification of the rhizosphere microflora by inoculation is a much discussed question, particularly the value of seed and root bacterization to improve crop yield, and not specifically to control disease (Brown, 1974). Ideally, the inoculant should be able to grow in the rhizosphere thus maintaining its population and giving maximum benefit to the plant, but in practice, the population often declines after a few weeks, being unable to compete successfully with the established root microflora. However, survival of the inoculant during the early stages of seedling growth may have considerable value because metabolites such as growth regulators produced at this stage will modify future plant development. Also, an inoculant that lyses, antagonizes or competes with a pathogen causing seedling diseases may suppress the pathogen sufficiently to decrease infection rate significantly.

The usual methods of introduction are by coating seeds with suspensions of the micro-organisms or by dipping roots of germinated seedlings into the suspension before transplanting to the growing medium. As a routine procedure seed inoculation has certain practical difficulties because bacteria applied directly to seed treated with fungicide are liable to be killed, or if coated seed is combine-drilled with mineral fertilizers the bacteria can be killed by osmotic shock (Brown, Burlingham & Jackson, 1964). To overcome these difficulties various peat preparations have been devised that can be sprinkled on the soil or seed at the time of sowing (Rubenchik, 1963; Ridge & Rovira, 1968), but they have varied in their effectiveness because of death of the bacteria. Peat blocks with added nutrients formulated according to crop requirements and wetted with a culture of the inoculant have proved very effective for vegetable crops whose seedlings are raised in them before transplanting (Mishustin, 1966). After inoculation the micro-organisms usually spread from the seed and colonize young roots (Jackson & Brown, 1966) and sometimes the introduced bacteria survive in the rhizosphere until the plants are harvested (Brown, Burlingham & Jackson, 1962).

Using certain bacteria, notably *Azotobacter chroococcum,* yields of crops have been increased, particularly plants like cabbage and lettuce that are harvested in full leaf, or carrot where the storage organ is the crop. Cereals sometimes respond, but increases in yield are usually $< 10\%$ (Brown, 1974). The practice of this type of inoculation started in the Soviet Union about 1927 when mineral fertilizers were unavailable, and by 1958 $c.10$ million hectares were treated. *Azotobacter chroococcum* was selected because it fixed atmospheric nitrogen and might therefore add extra nitrogen to the soil. A strain of *Bacillus megaterium* was also used because it mineralized organic phosphorus compounds rapidly in culture media. Crop growth is improved, but not because of extra nitrogen or phosphorus. Other bacteria belonging to *Pseudomonas, Clostridium* and *Bacillus* genera also improve growth. The inoculants are more effective when used in conjuction with mineral and organic fertilizers and in conditions optimum for plant growth. All the inoculants affect growth similarly, frequently increasing germination rates, elongating stems, enlarging leaves and causing earlier flowering and fruiting, effects indicating a common cause related to growth regulators.

All the inoculants produce growth regulators belonging to the auxin and gibberellin groups and some produce cytokinins, and the fact that on application to seed or seedling the inoculum contains enough growth regulator to alter subsequent plant development precludes the necessity of continuous multiplication in the rhizosphere. At the critical stage of seedling growth the inoculants are usually dominant members of the rhizosphere microflora and may continuously be producing more growth regulator which is absorbed by the roots. The magnitude of the plant response will be influenced by the amount of hormone produced and environmental conditions and variations in these may explain why responses cannot be predicted reliably.

Manipulating the rhizosphere microflora by inoculation specifically for disease control has met with limited success in natural soil, mainly because the inoculants are rapidly killed. *Bacillus subtilis* has proved a useful organism. It gave some control of pigeon-pea wilt when a suspension mixed with molasses and groundnut cake was applied to the seed, the substrates possibly enhanced antibiotic production in the rhizosphere (Singh, Vasudeva & Bajaj, 1965). This bacterium also controlled seedling blight of maize when coated on the kernels. Both pericarp and roots were protected against the pathogen and control was more effective than by using captan or thiram (Chang & Kommendahl, 1968). *Bacillus subtilis* was used to inoculate barley, wheat and oat seed sown in areas in a field affected by 'bare patch' caused by *Rhizoctonia solani, Pythium* sp. and *Fusarium* sp. Wheat and oats showed increased tillering and dry weight, and barley and oats gave 9% increase in yield and matured 2 weeks earlier than plants from untreated seed. The suspected pathogens were still present but fewer pseudomonads were in the patches; this may be the effect of fewer roots in the patch and therefore less nutrient for the pseudomonads (Price *et al.,* 1971). In other tests in which water suspensions of *B. subtilis* and a *Streptomyces* sp. were applied to oat seeds the grain yield was increased by 40–45%, and yield of carrots was increased by 48% after pelleting the seed with *B. subtilis* (Merriman *et al.,* 1974). In these successful experiments the isolates were made directly from the surface mycelium of the pathogen or from selected soil, and screened in agar plates for antibiotic activity against 6 or more distinct plant pathogenic fungi, but the authors considered that plant growth responses might be due to factors other than biological control of root pathogens.

Arthrobacter spp. were also used as a soil drench and seed inoculant to control damping-off of tomato (Mitchell & Hurwitz, 1965), *Pythium* mycelium was lysed only by the seed inoculant.

Inoculation of seedlings or cuttings grown in specialized propagation media has been very successful; for example, the control of *Fusarium* wilt of carnation cuttings by *B. subtilis* (Aldrich & Baker, 1970), and of *Pythium ultimum* on antirrhinum (Broadbent, Baker & Waterworth, 1971). In other experiments *B. subtilis* controlled *Rhizoctonia solani* by lysing the mycelium, the degree of protection varying with soil and with strains of fungi and bacteria (Olsen & Baker, 1968).

Extensive commercial trials were made in a New South Wales nursery to determine the feasibility of inoculating treated soil with bacteria to increase the growth rate of bedding-plant seedlings (Baker & Cook, 1974). Seed germination rate and weight of seedlings was increased for a number of plants by 2 strains of *B. subtilis* although there was marked specificity in the plant-bacteria response, inoculation sometimes causing decreases rather than increases.

It is of considerable interest that many of the bacteria giving control of disease do so with accompanying changes in growth of plants, suggesting that growth regulators may be involved in the process. There is some evidence that regulators may alter the expression of disease either by regulating parasite growth and toxin production or by modifying the host response (Davis & Dimond, 1953; Volken, 1972).

6. Changes in the Microflora Resulting from Crop Management

In these days of intensive agriculture many farmers are growing the same crops on the same land over a period of years. Inevitably with a continuous supply of similar substrate the microflora will be selected to one best able to use that substrate, and this may also mean a selection for organisms pathogenic for the particular crop. In some cases disease levels become so great that it is no longer possible to grow the crop, but in others, disease is at first severe but then becomes less; a natural biological control takes place and the soil is pathogen-suppressive.

There are examples of a decline in the intensity of disease caused by 3 important plant pathogens. Menzies (1959) reported that soils long used for potato production become resistant to *Streptomyces scabies* and production of a transferable, heat labile factor was stimulated by adding alfalfa meal to the soil. Burke (1965) found that soils developed an adaptive resistance to root rot of beans caused by *Fusarium*; in these soils the chlamydospores were less resistant and less persistant than in virgin soils. The factor causing this change was thermolabile, but not transferable. The third decline observed in disease, and the most studied, is that of take-all caused by *Gaeumannomyces graminis* var. *tritici* (*Ophiobolus graminis*) (Fig. 1). Take-all disease is a major problem in cereal growing and is unalleviated by plant protection chemicals or resistant varieties, take-all decline makes consecutive cereal cropping practicable and may offer a means of control.

The cause of take-all decline is unknown but several people have suggested

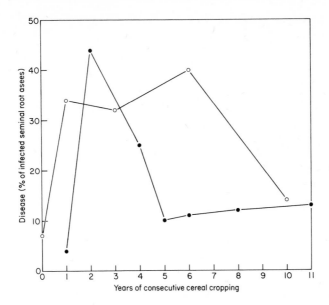

Fig. 1. The effect of consecutive cereal cropping on the amount of take-all disease in soil of Little Knott field, Rothamsted. Samples were taken from plots that had carried cereals for different numbers of consecutive years. ○, samples taken on 6 October 1970; ●, samples taken on 10 September 1971.

reasonable mechanisms; Shipton (1972) supports a popular hypothesis, owing much to the work of Gerlagh (1968), suggesting that a 'specific' antagonism operates from soil micro-organisms developing in response to the increasing population of *G. graminis*. Pope & Jackson (1973) and Vojinović (1973) suggest that the microflora in decline soils affects the growth rate and tropic response of the pathogen's hyphae to wheat roots. Other hypotheses rely on mycoparasitism and virus infection of *G. graminis*, but lack proof. Brown, Hornby & Pearson (1973) think the decline may operate through the changes in the microflora modifying the root environment nutritionally, possibly by changes in the $NO_3^- -N : NH_4^+ -N$ ratio; certainly changes in host and pathogen nutrition modify the infection level (Brown, unpublished). Continued work on this problem may eventually link together the separate observations to provide the answer and suggest how it may be used in agriculture.

7. Conclusions

Growth of plants in their natural environment is governed by many factors, one of which is the composition of the root microflora with all its capacity for good and bad. Manipulating this microflora is difficult for we are trying to change an integrated population in equilibrium and we may cause a series of changes with results that are unforeseen. The aim should be to increase the population that most benefits the plant. Further research into biological control could profitably explore the examples already given us in nature. In attempts to manipulate the microflora by inoculation more success may be achieved with use of mixed inocula of organisms that antagonize, lyse and compete with the pathogen, together with those that may alter the root environment by their metabolic activity.

The idea of manipulating the microflora should be approached with confidence, even if we do not fully understand the complexities of the situation; if we wait for total knowledge we will not achieve success for many decades. Rather let us try to make a particular system work, then study it carefully to reach perfection.

8. References

ALDRICH, J. & BAKER, R. (1970). Biological control of *Fusarium roseum* f. sp. *dianthi* by *Bacillus subtilis*. *Pl. Dis. Reptr* **54**, 446.

ALEXANDER, M. (1961). *Introduction to Soil Microbiology*. New York & London: John Wiley & Sons Inc.

BAKER, K. F. & COOK, R. J. (1974). *Biological Control of Plant Pathogens*. p. 97. San Francisco: W. H. Freeman & Co.

BARBER, D. A. (1968). Micro-organisms and the inorganic nutrition of higher plants. *Ann. Rev. Pl. Physiol.* **19**, 71.

BAREA, J. M. & BROWN, M. E. (1974). Effects on plant growth produced by *Azotobacter paspali* related to synthesis of plant growth regulating substances. *J. appl. Bact.* **37**, 583.

BLONDEAU, R. (1970). Production d'une substance de type cytokinine par des *Arthrobacter* d'origine rhizosphère. *C. r. hebd. Séanc. Acad. Sci., Paris* **270D**, 3158.

BOWEN, G. D. & ROVIRA, A. D. (1961). The effects of micro-organisms on plant growth. 1. Development of roots and root hairs in sand and agar. *Pl. Soil* **15**, 166.

BOWEN, G. D. & ROVIRA, A. D. (1973). Are modelling approaches useful in rhizosphere biology? In *Modern Methods in the Study of Microbial Ecology*. Ed. T. Rosswall. Stockholm: Swedish National Research Council.

BROADBENT, P., BAKER, K. F. & WATERWORTH, Y. (1971). Bacteria and actino-mycetes antagonistic to fungal root pathogens in Australian soils. *Aust. J. biol. Sci.* **24**, 925.

BROWN, M. E. (1972). Plant growth substances produced by micro-organisms of soil and rhizosphere. *J. appl. Bact.* **35**, 443.

BROWN, M. E. (1974). Seed and root bacterization. *A. Rev. Phytopath.* **72**, 181.

BROWN, M. E., BURLINGHAM, S. K. & JACKSON, R. M. (1962). Studies on *Azotobacter* species in soil. II. Populations of Azotobacter in the rhizosphere and effects of artificial inoculation. *Pl. Soil.* **17**, 320.

BROWN, M. E., BURLINGHAM, S. K. & JACKSON, R. M. (1964). Studies on *Azotobacter* species in soil. III. Effects of artificial inoculation on crop yields. *Pl. Soil.* **20**, 194.

BROWN, M. E., HORNBY, D. & PEARSON, V. (1973). Microbial populations and nitrogen in soil growing consecutive cereal crops infected with take-all. *J. Soil Sci.* **24**, 296.

BROWN, M. E., JACKSON, R. M. & BURLINGHAM, S. K. (1968). Effects produced on tomato plants, *Lycopersicum esculentum,* by seed or root treatment with gibberellic acid and indolyl-3-acetic acid. *J. exp. Bot.* **19**, 544.

BURKE, D. W. (1965). *Fusarium* root rot of beans and behaviour of the pathogen in different soils. *Phytopathology* 55, 1122.

CHANG, I. & KOMMENDAHL, T. (1968). Biological control of seedling blight of corn by coating kernels with antagonistic micro-organisms. *Phytopathology* 58, 1395.

COPPOLA, S., PERCUOCO, G., ZOÏNA, A. & PICCI, G. (1971). Citochinine in germi terricoli e relativo significato nei rapporti piante microorganismi. *Annali Microbiol.* 21, 45.

DAVIS, D. & DIMOND, A. E. (1953). Inducing disease resistance with plant growth regulators. *Phytopathology* 43, 137.

FOSTER, R. C. & ROVIRA, A. D. (1973). The rhizosphere of wheat roots studied by electron microscopy of ultra thin sections. In *Modern Methods in the Study of Microbial Ecology*. Ed. T. Rosswall. Stockholm: Swedish National Research Council.

GERLAGH, M. (1968). Introduction of *Ophiobolus graminis* into new polders and its decline. *Meded. Lab. Fytopathologie, Wageningen.* No. 241.

GORING, C. A. I. & CLARK, F. E. (1948). Influence of crop growth and mineralization of nitrogen in the soil. *Proc. Soil Sci. Soc. Am.* 13, 261.

GREAVES, M. P. & WEBLEY, D. M. (1965). A study of the breakdown of organic phosphates by micro-organisms from the root region of certain pasture grasses. *J. appl. Bact.* 28, 454.

GREAVES, M. P. & WEBLEY, D. M. (1969). The hydrolysis of *myo*-inositol hexaphosphate by soil micro-organisms. *Soil Biol. Biochem.* 1, 37.

GRUEN, H. E. (1959). Auxins and fungi. *Ann. Rev. Pl. Physiol.* 10, 405.

HELGESON, J. P. & LEONARD, N. J. (1966). Cytokinins: identification of compounds isolated from *Corynebacterium fascians. Proc. natn. Acad. Sci.* U.S.A. 56, 60.

JACKSON, R. M. & BROWN, M. E. (1966). Behaviour of *Azotobacter chroococcum* introduced into the plant rhizosphere. *Annls Inst. Pasteur, Paris.* 111, suppl. 3, 103.

JOHNSTON, W. B. & OLSEN, R. A. (1972). Dissolution of fluorapatite by plant roots. *Soil Sci.* 114, 29.

KATZNELSON, H., LOCHHEAD, A. G. & TIMONIN, M. I. (1948). Soil micro-organisms and the rhizosphere. *Bot. Rev.* 14, 543.

KATZNELSON, H., ROUATT, J. W. & PAYNE, T. M. B. (1955). The liberation of amino acids and reducing compounds by plant roots. *Pl. Soil* 7, 35.

LIBBERT, E. & MANTEUFFEL, R. (1970). Interactions between plants and epiphytic bacteria regarding their auxin metabolism. VII. The influence of the epiphytic bacteria on the amount of diffusable auxin from corn coleoptiles. *Physiol. Plantarum* 23, 93.

LOCKWOOD, J. L. (1964). Soil fungistasis. *Ann. Rev. Phytopath.* 2, 341.

LOUTIT, M. W. & BROOKS, R. R. (1970). Rhizosphere organisms and molybdenum concentrations in plants. *Soil Biol. Biochem.* 2, 131.

LOUTIT, M. W., LOUTIT, J. S. & BROOKS, R. R. (1967a). Differences in molybdenum uptake by micro-organisms from the rhizosphere of *Raphanus sativus* L. grown in two soils of similar origin. *Pl. Soil* 27, 335.

LOUTIT, M. W., MALTHUS, R. S. & LOUTIT, J. S. (1967b). The effect of soil micro-organisms on the concentration of molybdenum in the radish (*Raphanus sativus* L.) variety 'White Icicle'. *N.Z. J. agric. Res.* 11, 420.

MENZIES, J. D. (1959). Occurrence and transfer of a biological factor that suppresses potato scab. *Phytopathology* 49, 648.

MERRIMAN, P. R., PRICE, R. D., KOLLMORGEN, J. F., PIGGOTT, T. & RIDGE, E. H. (1974). Effect of seed inoculation with *Bacillus subtilis* and *Streptomyces griseus* on the growth of cereals and carrots. *Aust. J. agr. Res.* 25, 219.

MISHUSTIN, E. N. (1966). Action d'*Azotobacter* sur les végétaux supérieurs. *Annls Inst. Pasteur, Paris* 111, suppl. 3, 121.

MISHUSTIN, E. N. (1972). Processes microbiologiques mobilisant les composés du phosphore dans le sol. *Revue Ecol. & Biol. Sol.* (Fr.) 9, 521.

MITCHELL, R. & HURWITZ, E. (1965). Suppression of *Pythium debaryanum* by lytic rhizosphere bacteria. *Phytopathology* 55, 156.

NAIR, P. K. R. & TALIBUDEEN, O. (1973). Dynamics of K and NO_3 concentrations in the root zone of winter wheat at Broadbalk using specific-ion electrodes. *J. agric. Sci., Camb.* **81**, 327.

OLSEN, C. M. & BAKER, K. F. (1968). Selective heat treatment of soil, and its effect on the inhibition of *Rhizoctonia solani* by *Bacillus subtilis. Phytopathology* **58**, 79.

PARKINSON, D. (1967). Soil micro-organisms and plant roots. In *Soil Biology.* Eds A. Burges & F. Raw. London: Academic Press.

PHILLIPS, D. A. & TORREY, J. G. (1970). Cytokinin production by *Rhizobium japonicum. Physiologia Pl.* **23**, 1057.

PINCKARD, J. A. (1970). Microbiological antagonism encouraged by the monosodium salt of hexachlorophene. *Phytopathology* **60**, 1308. (Abstr.)

POPE, A. M. S. & JACKSON, R. M. (1973). Effects of wheat field soil on inocula of *Gaeumannomyces graminis* (Sacc.) Arx and Olivier var. *tritici* J. Walker in relation to take-all decline. *Soil Biol. Biochem.* **5**, 881.

PRICE, R. D., BAKER, K. F., BROADBENT, P. & RIDGE, E. H. (1971). Effect on wheat plants of a soil or seed application of *Bacillus subtilis* either with or without the presence of *Rhizoctonia solani. Proc. Aust. Conf. Soil Biol.* 50.

RIDGE, E. H. & ROVIRA, A. D. (1968). Microbial inoculation of wheat. *Trans. 9th Intern. Congr. Soil Sci.* III. 473.

ROMANOW, I., CHALVIGNAC, M. A. & POCHON, J. (1969). Recherches sur la production d'une substance cytokinique par *Agrobacterium tumefaciens* (Smith et Town) Conn. *Annls Inst. Pasteur, Paris* **117**, 58.

ROVIRA, A. D. (1956). A study of the development of the root surface microflora during the initial stages of plant growth. *J. appl. Bact.* **19**, 72.

ROVIRA, A. D. (1959). Root excretions in relation to the rhizosphere effect. IV. Influence of plant species, age of plants, light, temperature and calcium nutrition on exudation. *Pl. Soil* **11**, 53.

ROVIRA, A. D. (1965*a*). Interactions between plant roots and soil micro-organisms. *Ann. Rev. Microbiol.* **19**, 241.

ROVIRA, A. D. (1965*b*). Plant root exudates and their influence upon soil micro-organisms. In *Ecology of Soil-Borne Plant Pathogens – Prelude to Biological Control.* Eds K. F. Baker & W. C. Snyder. Berkeley: University of California Press.

ROVIRA, A. D. (1969). Plant root exudates. *Bot. Rev.* **35**, 35.

ROVIRA, A. D. & CAMPBELL, R. (1974). Scanning electron microscopy of micro-organisms on the roots of wheat. *Microbial Ecology* **1**, 15.

RUBENCHIK, L. I. (1963). Azotobacter *and Its Use in Agriculture.* Jerusalem: Israel Program for Scientific Translations.

SAUCHELLI, V. (1965). *Phosphates in Agriculture.* Reinhold Publishing Corporation, New York & London: Chapman & Hall.

SCHROTH, M. N. & HILDEBRAND, D. C. (1964). Influence of plant exudates on root infecting fungi. *Ann. Rev. Phytopath.* **2**, 101.

SCOTT, T. K. (1972). Auxins and roots. *Ann. Rev. Physiol.* **23**, 235.

SHIPTON, P. J. (1972). Take-all in spring-sown cereals under continuous cultivation: disease progress and decline in relation to crop succession and nitrogen. *Ann. appl. Biol.* **71**, 33.

SINGH, P., VASUDEVA, R. S. & BAJAJ, B. S. (1965). Seed bacterization and biological activity of bulbiformin. *Ann. appl. Biol.* **55**, 89.

SKOOG, F. & ARMSTRONG, D. J. (1970). Cytokinins. *Ann. rev. Pl. Physiol.* **21**, 359.

STANKOVA-OPOCENSKA, E. & DEKKER, J. (1970). Indirect effect of 6-azauracil on *Pythium debaryanum. Neth. J. Pl. Path.* **76**, 152.

STARKEY, R. L. (1958). Inter-relations between micro-organisms and plant roots in the rhizosphere. *Bact. Rev.* **32**, 154.

STREET, H. E. (1969). Growth in organized and unorganized systems. Knowledge gained by culture of organs and tissue explants. In *Plant Physiology V*B. (Analysis of Growth: The Responses of Cells and Tissues in Culture). Ed. F. C. Steward. New York & London: Academic Press.

STRZELCZYK, E., KAMPERT, M. & KRZYSKO, K. (1971). Production of inhibitors of auxin and gibberellin induced growth of plants by *Arthrobacter pascens*. *Act. microbiol. pol.* series A. **111**, 85.

SWABY, R. J. (1942). Stimulation of plant growth by organic matter. *J. Aus. Inst. agric. Sci.* **8**, 156.

SWABY, R. J. & SPERBER, J. (1959). Phosphate dissolving micro-organisms in the rhizosphere of legumes. In *Nutrition of Legumes*. Ed. E. D. Hallsworth. London: Butterworths.

THIMANN, K. K. (1972). The Natural Plant Hormones. In *Plant Physiology*. *V*IB. (Physiology of Development. The Hormones). Ed. F. C. Steward. London: Academic Press.

VANČURA, V. & HANZLÍKOVÁ, A. (1972). Root exudates of plants. IV. Differences in chemical composition of seed and seedling exudates. *Pl. Soil* **36**, 271.

VOJINOVIĆ, Z. D. (1973). The influence of micro-organisms following *Ophiobolus graminis* Sacc. on its further pathogenicity. *Org. Eur. Med. Prot. Plantes Bull.* **9**, 91.

VOLKEN, P. (1972). Quelques aspects des relations hôte-parasite en fonction de traitements à l'acide indol-acétique et à l'acide gibberellique. *Phytopathol. Z.* **75**, 163.

VRANÝ, J. (1965). Effect of foliar application on the rhizosphere microflora. In *Plant Microbes Relationships*. Eds J. Macura & V. Vančura. Prague: Czechoslovak Academy of Science.

VRANÝ, J., VANČURA, V. & MACURA, J. (1962). The effect of foliar application of some readily metabolized substances, growth regulators and antibiotics on rhizosphere microflora. *Fol. microbiol.* **7**, 61.

WEINHOLD, A. R. & BOWMAN, T. (1968). Selective inhibition of the potato scab pathogen by antogonistic bacteria and substrate influence on antibiotic production. *Pl. Soil* **27**, 12.

ZAGALLO, A. C. & KATZNELSON, H. (1957). Metabolic activity of bacterial isolates from wheat rhizosphere and control soil. *J. Bact.* **73**, 760.

Bacterial Diseases of Farmed Fishes

R. J. ROBERTS

*Unit of Aquatic Pathobiology, University of Stirling,
Stirling FK9 4LA, Scotland*

CONTENTS

1. Introduction

ALTHOUGH AQUACULTURE has been practised for some 2000 years, it is only in the last 20 years that it has developed from an extensive, low capital industry involving a small number of freshwater species of fish to become a significant source of high quality protein food. The range of species of fish which are now farmed has widened considerably and current developments, especially in marine culture, may well allow commercial farm production of most of the sea fish species currently obtained from the wild, although it is unlikely that under present circumstances farmed fish will replace wild caught species completely.

There are many factors, biological and economic, which affect the feasibility of fish farming, but one factor above all others which can destroy the economic viability of a farm is infectious disease (Roberts & Shepherd, 1975).

In any intensive farming system the relationship between the microbes of the environment and the defensive mechanisms of the species being farmed is critical and is usually dependent on diverse factors such as stocking density, husbandry methods and quality of stockmanship. In aquaculture these factors all apply, but one of the most important features of this inter-relationship is the temperature of the environment.

Fish are ectotherms (poikilotherms) and so do not have the sophisticated temperature control mechanism possessed by higher vertebrates which allows control of body temperature irrespective of ambient temperature. Consequently the body temperature of the fish is virtually that of the surroundings and varies as the water temperature varies.

The aquatic bacteria which are responsible for causing diseases in farmed fish have optimum temperature ranges for growth. Similarly the various fish species have optimum temperature ranges over which their defences against bacterial invasion are most efficient. Outbreaks of disease, therefore, are usually associated with sudden temperature change, which swings the balance in favour of the more rapidly acclimatizing bacteria, or else occurs at the extremes of the temperature range for the fish species concerned, especially where there are also other environmental stresses such as pollution, overcrowding or oxygen deficiency.

Table 1 defines the major diseases, with their clinical features, aetiological agents and pre-disposing causes, but in many cases the aetiology is complex. The Gram negative bacteria are the most common cause of severe septicaemic disease on the fish farm, with the *Aeromonas, Pseudomonas* and *Vibrio* genera being most frequently implicated. Great care is necessary in reaching a diagnosis of primary Gram negative septicaemia because in moribund fish, especially at higher temperatures, organisms from the gut can invade the tissues and produce a terminal bacteraemia which masks the primary cause of mortality.

There is only a limited number of bacterial groups associated with serious diseases in aquaculture. Almost all of these are normally associated with soil or water. The major species of bacteria which are responsible for disease in farmed fish are listed in Table 2 along with their principal diagnostic features. In many cases, notably with the Bacterial Kidney Disease organism (*Corynebacterium*) and the Red Mouth bacterium, culture and isolation difficulties are such that a full description and accepted nomenclature is not available.

2. The Bacterial Diseases

(a) *Furunculosis*

The most important bacterial disease in most salmonid farming countries is furunculosis, caused by the Gram negative *Aeromonas salmonicida* (McGraw, 1952). It was the subject of an official enquiry by the British Government in the 1920s and 1930s, because of the severe losses it was causing in Scottish East coast rivers and the 3 official reports (Mackie *et al.*, 1930, 1933, 1935) provided the basis for much of our present knowledge. The organism is widely distributed in river systems and is thought to survive between epizootics, within the kidney and intestine of carrier fish. It is released when the carrier is sufficiently stressed by environmental factors, notably temperature or hypoxia, allowing dissemination of infection through the population. In older fish the bacteria usually localize under the skin, causing the production of the characteristic 'furuncles'

Table 1

Principal bacterial diseases of fish culture

Disease	Causative bacterium	Environmental factors favouring disease	Clinico-pathological features
Furunculosis	*Aeromonas salmonicida*	Higher temperature, low oxygen. Carrier fish.	Sudden death in young fish. Furuncles + haemorrhages in older fish.
Haemorrhagic septicaemia	*Pseudomonas fluorescens* *Aeromonas hydrophila*	High levels of organic matter in water. High temperatures.	Haemorrhages, liquefactive necrosis of kidney.
Vibriosis	*Vibrio anguillarum*	High temperatures, skin trauma.	Red blotches on skin, deep muscular haemorrhages. Anaemia. Necrosis of haemopoietic and muscle tissue.
Myxobacterioses (Columnaris disease, cold-water disease, myxobacterial gill disease)	*Flexibacter columnaris* *Flexibacter psychrophilia* *Flexibacter* spp.	High temperatures/low oxygen levels. Low temperatures/detritus. Food debris in small fish.	Whitish lesions on head. Necrosis of extremities. Gasping. Swollen head.
Tuberculosis	*Mycobacterium* spp.	Feeding infected material.	Cachexia with whitish lesions.
Corynebacterial kidney disease	*Corynebacterium* spp.	Low pH value of water, carrier fish.	Punctate haemorrhages. White granulomatous lesions in kidney, muscle and liver. Diphtheritic membrane formation on organs at low temperatures.
Nocardiosis	*Nocardia asteroides*		Swollen mouth or abdomen.
Streptomycoses	*Streptomyces salmonicida*		White granulomas on intestinal wall.
Red Mouth	'R M' bacterium		Haemorrhagic stomatitis

Table 2
Principal bacteria responsible for disease in fish culture

Species	Staining	Morphology	Oxidase reaction	Pigment production Diffusible	Pigment production Non-diffusible	Motility	Inhibition by Novobiocin	Inhibition by 0/129 Vibriostat	Glucose fermentation	'O–F medium'	Oxygen requirement
Aeromonas salmonicida	Gram −ve	bacillus	+	Brown (can be −ve)	−	−	−	−	acid + gas	fermentative	facultatively anaerobic
A. hydrophila	Gram −ve	bacillus	+	− rarely + (Brown)	−	+ polar flagella	−	−	acid + gas	fermentative	aerobic + facultatively anaerobic
Pseudomonas fluorescens	Gram	bacillus	+	green fluorescent	−	+ polar flagella	−	−	acid	oxidative	aerobic
Vibrio anguillarum	Gram −ve	vibrion	+	−	−	+ polar flagella	+	+	acid	fermentative	facultatively anaerobic
Flexibacter psychrophila	Gram −ve	long slender bacillus	+	−	yellow	−	−	−			aerobic
Flexibacter columnaris	Gram −ve	long slender bacillus	+	−	yellow	−	−	−			aerobic
'Red Mouth' bacterium (Enterobacteraceae)	Gram −ve	bacillus	−	−	−	+ peritrichous flagella	−	−	acid	fermentative	facultatively anaerobic
Corynebacterium sp.	Gram +ve	diphtheroid bacillus									
Mycobacterium spp.	acid fast	bacillus									aerobic
Nocardia asteroides	Gram +ve, occasionally acid-fast	mycelium									aerobic
Streptomyces spp.	Gram +ve	mycelia with conidia									

which are the main clinical feature and are a major source of infection when they rupture. *Aeromonas salmonicida* often causes acute mortality with no obvious clinical features apart from darkening of the skin in young salmonids. Internally, however, microscopic lesions can be found in the heart and kidney and *A. salmonicida* can be isolated in pure culture from the blood and tissues. *Aeromonas salmonicida* produces a leucocidin and also has a lipopolysaccharide moiety which is present in smaller amount in avirulent strains (Klontz, Yasutake & Ross, 1966; Anderson, 1974). Because of its significance to salmonid farmers, a number of attempts at vaccine production has been made, using chloroform-killed (Duff, 1942), formalinized (Krantz, Reddecliff & Heist, 1964) or alum-precipitated (Klontz & Anderson, 1970) antigens. No vaccine is yet available commercially, but protective serum antibody levels have been produced in experimental studies.

(b) *Haemorrhagic septicaemia*

Aeromonas hydrophila (including *A. liquefaciens*), other aeromonads and *Pseudomonas* spp. are micro-organisms which occur widely in waters with a heavy organic load. Unlike those caused by *A. salmonicida,* epizootics are not dependent on carrier fish, but fatal septicaemias can almost invariably be related to some other stress on the fish such as high temperature, sexual maturity (Thorpe & Roberts, 1972) or low oxygen levels. Affected fish show a severe haemorrhagic syndrome with blotchy haemorrhages of the skin and base of fins and, internally, liquefactive necrosis of the spleen and kidney and haemorrhagic enteritis. Outbreaks occur characteristically in high summer, when oxygen levels and water levels are low, or, especially in cyprinid fish, in spring or autumn. A disease of cyprinids called 'Carp Dropsy' characterized by ascites, exophthalmos and skin ulceration, was for long considered to be caused by *A. liquefaciens* (Schaperclaus, 1954), but recent work, notably that of Fijan (1972) has shown that Carp Dropsy is a very complex disease with a rhabdovirus, (*Rhabdovirus carpei*), a *Brucella*-like micro-organism and a number of other factors contributing to a situation where, although aeromonads are a major cause of death of the fish they are unlikely to be the primary aetiological agent of the disease.

(c) *Vibriosis*

Vibriosis is caused by *Vibrio anguillarum* and is a particularly severe septicaemic disease of marine fishes. The organism is usually halophilic and has many similarities to *V. cholerae*. The disease is characterized by deep focal necrotizing myositis and subdermal haemorrhages. Moribund fish also show a swollen spleen and kidney and, in longer standing cases, haemorrhagic stomatitis and exophthalmos. The bacterium produces a haemolysin which causes severe anaemia in those fish which survive the severe acute stage. *Vibrio anguillarum* can be isolated from a number of marine invertebrates and wild fish (Håstein & Holt, 1972) and the main predisposing cause of farm outbreaks is usually rise of temperature to 15° or above, although outbreaks have occurred at lower temperatures. Because of its significance to marine aquaculture, considerable efforts have been made to develop a suitable oral vaccine. Some success has been achieved with a dried sonicate of *V. anguillarum* (Fryer, Nelson & Garrison, 1972), but it has not yet been proved commercially. A number of as yet uncharacterized *Pasteurella*-like organisms also cause identical syndromes to vibriosis in many species of marine fishes (Håstein T. pers. comm.; Snieszko *et al.*, 1964).

(d) *Myxobacterial infection*

Myxobacteria or slime bacteria are ubiquitous in the aquatic environment, but are only weakly invasive. However, when they are presented with a lowering of the resistance of the integument by environmental stress such as high temperature, anoxia, accumulation of ammonia or detritus, or skin changes associated with precocious sexual maturity, they are capable of invading the skin. In young fish myxobacteria may grow on food particles lodged in the gill and by extending from here to the gill tissue proper, cause the condition known as 'bacterial gill disease'. Columnaris disease, caused by the myxobacterium *Flexibacter columnaris* occurs in a wide variety of freshwater species, almost invariably in association with high temperature stress.

(e) *Tuberculosis and other granulomata*

Tuberculosis is widespread in fishes but, although heavy losses occur in aquaria, this is rarely the case in the farm situation unless the fish are fed infected fish offals (Ross, 1963). Infections with other members of the Actinomycetales are infrequent but both *Streptomyces salmonicida* (Rucker, 1949) and *Nocardia asteroides* (Wolke & Meade, 1974) have been associated with losses in Pacific salmon culture.

(f) *Corynebacterial kidney disease*

An unnamed *Corynebacterium* is one of the most dangerous pathogens of salmonid culture in a number of countries. Primarily associated with water of low pH value, it proliferates slowly in the interstitium of the kidney, exciting a granulomatous response by the host, which eventually involves most of the kidney and frequently the spleen, liver and skeletal muscle also. The micro-organism is very small and readily confused with melanin granules in Gram stained smears of spleen or kidney. The organism is extremely fastidious in its growth requirements, growing very slowly even on the blood-cysteine medium specially designed for this purpose (Ordal & Earp, 1956). In adult Atlantic salmon, at lower temperatures, it causes a slightly different syndrome with diphtheresis of the peritoneal surfaces, often with extensive petechiation of the abdominal wall and viscera, although at higher temperatures a syndrome similar to that described for rainbow trout occurs (Smith, 1964). Corynebacterial kidney disease is of great economic significance and because of its insidious nature, heavy losses can result. Detection of carrier fish is all important and serological tests, coupled with ready isolation methods are required in order to allow efforts to eradicate it.

3. Acknowledgement

I would like to thank Miss Margaret Hendrie for her critical review of the paper and correction of several taxonomical inexactitudes.

4. References

ANDERSON, D. P. (1974). *Diseases of Fishes. IV Immunology.* Neptune, N. J.: TFH Publications.

DUFF, D. (1942). Oral immunization of trout against *Bacterium salmonicida. J. Immunol.* **44,** 87.

FIJAN, N. (1972). Infectious dropsy of carp — A disease complex. *Symp. Zool. Soc. Lond.* **30,** 39.

FRYER, J. L., NELSON, J. S. & GARRISON, R. L. (1972). Vibriosis in fish. *Progress in Fishery and Food Science* **5,** 129.

HÅSTEIN, T. & HOLT, G. (1972). The occurrence of Vibrio Disease in wild Norwegian fish. *J. Fish Biol.* **4,** 33.

KLONTZ, G. W. & ANDERSON, D. P. (1970). Oral immunization of salmonids, a review. *Am. Fish Soc. Spec. Publ.* **5,** 16.

KLONTZ, G. W., YASUTAKE, W. T. & ROSS, A. J. (1966). Bacterial diseases of Salmonidae in the Western United States: Pathogenesis of Furunculosis in rainbow trout. *Am. J. Vet. Res.* **27,** 1455.

KRANTZ, G. E., REDDECLIFF, J. M. & HEIST, C. E. (1964). Immune response of trout to *Aeromonas salmonicida. Prog. Fish Cult.* **26,** 3.

MACKIE, T. S., ARKWRIGHT, J. A., PRYCE-TENNANT, T. E., MOTTRAM, J. C., JOHNSTON, W. D. & MENZIES, W. J. M. (1930). *Furunculosis Committee First Interim Report.* Edinburgh: H.M.S.O.

MACKIE, T. S., ARKWRIGHT, J. A., PRYCE-TENNANT, T. E., MOTTRAM, J. C., JOHNSTON, W. D. & MENZIES, W. J. M. (1933). *Furunculosis Committee Second Interim Report.* Edinburgh: H.M.S.O.

MACKIE, T. S., ARKWRIGHT, J. A., PRYCE-TENNANT, T. E., MOTTRAM, J. C., JOHNSTON, W. D. & MENZIES, W. J. M. (1935). *Furunculosis Committee Final Report.* Edinburgh: H.M.S.O.

McGRAW, B. M. (1952). *Furunculosis of Fish.* Special Scientific Report: Fisheries No. 84. U.S. Fish & Wildlife Service.

ORDAL, E. J. & EARP, B. J. (1956). Cultivation and transmission of etiological agent of kidney disease in salmonid fishes. *J. exp. Biol. Med. Proc.* **92,** 85.

ROBERTS, R. J. & SHEPHERD, C. J. (1975). *A Handbook of Salmon & Trout Disease.* London: Fishing News (Books) Ltd.

ROSS, A. J. (1963). *Mycobacteria in Adult Salmonids.* Special Scientific Report. No. 462. U.S. Fisheries & Wildlife Service.

RUCKER, R. R. (1949). A streptomycete pathogenic to fish. *J. Bact.* **58,** 659.

SMITH, I. W. (1964). *The Occurrence and Pathology of Dee Disease.* D.A.F.S. Freshwater Fisheries Research Series 34. Edinburgh: H.M.S.O.

SCHAPERCLAUS, W. (1954). *Fischkrankheiten.* Berlin: Akademic Verlag.

SNIESZKO, S. F., BULLOCK, G. L., HOLLIS, E. & BOONE, J. G. (1964). *Pasteurella* from an epizootic of white perch (*Roccus americanus*) in Chesapeake Bay tide water areas. *J. Bact.* **88,** 1814.

THORPE, J. E. & ROBERTS, R. J. (1972). An aeromonad epidemic in the brown trout (*Salmo trutta*). *J. Fish Biol.* **4,** 441.

WOLKE, R. D. & MEADE, T. L. (1974). Nocardiosis in Chinook salmon. *J. Wildl. Dis.* **10,** 149.

Algal Lysing Agents of Freshwater Habitats

W. D. P. STEWART AND M. J. DAFT

*Department of Biological Sciences, University of Dundee,
Dundee DD1 4HN, Scotland*

CONTENTS

1. Introduction

THERE IS GOOD evidence from ecological studies carried out in countries throughout the world that increasing eutrophication is resulting in the development of excessive algal growths in many freshwater and inland coastal areas. Such growths develop when, as a result of change in the ecosystem (increased nutrient input in these particular cases), the original balance of the various components of the system breaks down and a new balance with heavy algal growths is established. In any aquatic ecosystem the components which contribute to such a natural balance are physical and biological. In this paper we would like to consider, in particular, a sub-group of the biological components — the viruses and bacteria (including actinomycetes) which occur in aquatic habitats as pathogens of the prokaryotic blue-green algae. The fungal pathogens of blue-green algae, studied in detail by Canter (Canter, 1972) will be mentioned only in passing. The various algal lysing agents are distributed widely in freshwater and marine habitats, and the available evidence suggests that they may play an important but as yet poorly understood role in regulating growth in aquatic environments.

2. Organisms

The organisms known to cause lysis of blue-green algae are listed in Table 1. Of the 4 major groups, the viruses and fungi are very specific in their host range, infecting only a few genera as do the LPP viruses (Padan & Shilo, 1973; Stewart & Daft, 1975) or even a single species as do certain fungi such as *Rhizophydium ubiquetum* on *Anabaena solitaria* (Canter, 1968). The actinomycetes and bacteria on the other hand usually (but not always, see Ensign, 1971) cause lysis of a wide range of blue-green algae and heterotrophic bacteria (Daft & Stewart, 1971) and some strains may also cause the lysis of eukaryotic algae (Stewart & Brown, 1970).

(a) *Viruses*

The possible importance of algal viruses in regulating growths of blue-green algae was recognized (Krauss, 1960) several years before the first virus of a pigmented blue-green alga was isolated. The latter virus, LPP-1, was isolated by Safferman & Morris (1963*a*) from a sewage oxidation pond and from work on this one organism, and the others listed in Table 1, has arisen a vast literature on viruses of blue-green algae (cyanophages). These viruses differ little from bacteriophages and actinophages. Although the LPP-group is widespread in distribution, the others have been found only rarely despite fairly extensive searches. The literature on cyanophages is detailed in several recent reviews (Brown, 1972; Padan & Shilo, 1973; Stewart & Daft, 1976).

Virus LPP-1 is typical of the LPP group (of which there are 2 serologically distinct sub-groups, LPP-1 and LPP-2) and has been studied most extensively. Our isolate (D-1) has an icosahedral head (58.6 ± 2 nm diam.) with a short non-contractile tail (20–22.5 nm long and 15 nm diam.) with a 6-fold axial symmetry. Viruses of the LPP-group resemble bacteriophages more closely than they do viruses of higher plants but nevertheless show no serological affinity with bacteriophages. The DNA is linear and double stranded, *c.* 13 nm long (Luftig & Haselkorn, 1967), the molecular weight is 27×10^6 daltons (Luftig & Haselkorn, 1967) and the G+C ratio is 55–57% (Goldstein & Bendet, 1967). Luftig & Haselkorn (1967) who carried out hybridization experiments found that there was little homology (0.25%) between the DNA of LPP-1 and that of *Plectonema boryanum* which indicates that this virus is unlikely to be very useful in transduction experiments. The virus is hydrated (0.37 g of water/g of virus) and is composed mainly of 2 proteins, one of mass 44 000 daltons and the other of mass 14 000 daltons.

Despite the discovery of LPP-1 in 1963 it was another 6 years before a second type of algal lysing virus was characterized (Safferman *et al.*, 1969*b*). This particle, SM-1, is effective against the unicellular algae *Synechococcus elongatus* and *Microcystis aeruginosa* NRC-1. Under the electron microscope the particles

Table 1

Pathogens of blue-green algae

Pathogen	Host range	Country of origin	Reference
Virus			
A-1(L)	*Anabaena variabilis*	Russia	Koz'yakov *et al.* (1972)
Ap-1*	*Aphanizomenon flos-aquae*	Sweden	Granhall (1972)
AR-1*	*Anabaenopsis raciborskii*	India	Singh & Singh (1967)
	Anabaenopsis circularis		
	Raphidiopsis indica		
AS-1	*Anacystis nidulans*	U.S.A.	Safferman *et al.* (1972)
	Synechococcus cedrorum		
AS-2*	*Anacystis nidulans*	Scotland	Daft & Stewart (unpub.)
	Synechococcus cedrorum		
	Synechococcus sp. (NRC-1)		
C-1*	*Cylindrospermum* sp.	India	Singh & Singh (1967)
LPP-1		U.S.A.	Safferman & Morris (1963*a*)
LPP-2		U.S.A.	Safferman *et al.* (1969*a*)
LPP-1D	*Lyngbya* spp.	Israel	Cannon *et al.* (1971)
LPP-1G	*Phormidium* spp.	Scotland	Padan *et al.* (1967)
D-1	*Plectonema* spp.		Daft *et al.* (1970)
P2*		India	Singh & Singh (1967)
P3*		India	Singh & Singh (1967)
P4*		India	Singh & Singh (1967)
N-1	*Nostoc muscorum*	U.S.A.	Adolph & Haselkorn (1971)
S-1	*Synechococcus* sp. (NRC-1)	U.S.A.	Adolph & Haselkorn (1973)
SM-1	*Synechococcus elongatus*	U.S.A.	Safferman *et al.* (1969*b*)
	Microcystis aeruginosa†		
Phage*	*Anabaena variabilis* strain 5	Sweden	Granhall & von Hofsten (1969)
Phage*	*Microcystis pulverea*	Russia	Rubenchyk *et al.* (1966)
Phage*	*Microcystis aeruginosa*	Russia	Goryushin & Chaplinskaya (1966)
	M. pulverea		
	M. incerta		
	M. muscicola		

* Full details of this cyanophage have not been published.
† Now classified as *Synechococcus* sp. NRC-1.

Table 1—*continued*

Pathogen	Host range	Country of origin	Reference
Bacterium			
Cellvibrio fulvus	*Anabaena inaequalis* *Chlorogloea fritschii* *Nodularia spumigena* *Nostoc punctiforme* *N. muscorum*	Sweden	Granhall & Berg (1972)
Flexibacter flexilis	*Anabaena cylindrica* *A. variabilis* *Nostoc punctiforme* *N. linckia* *Phormidium tadzchicicum* *P. luridum*	Russia	Gromov *et al.* (1972)
Myxococcus sp.	*Synechococcus cedrorum* *Lyngbya* sp.	U.S.A.	Wu *et al.* (1968)
Myxococcus fulvus *Myxococcus xanthus* *Myxococcus* sp. *Myxobacter* FP-1	*Nostoc muscorum* *Nostoc muscorum* *Nostoc muscorum* *Anacystis nidulans* *Coccochloris peniocystis* *Synechococcus cedrorum* *Nostoc* sp. *Plectonema boryanum* *Oscillatoria prolifera* *Spirulina platensis* *S. tenuis*	U.S.A. U.S.A. U.S.A. Israel	Stewart & Brown (1971) Stewart & Brown (1971) Stewart & Brown (1971) Shilo (1970)
Myxobacter CP-1	*Anabaena* (18 strains) *Anabaenopsis circularis* *Anacystis nidulans* *Aphanizomenon* (5 strains) *Coelosphaerium* sp.	Scotland	Daft & Stewart (1971, 1973) Daft *et al.* (1973, 1975)

Table 1 *—continued*

Pathogen	Host range	Country of origin	Reference
Myxobacter 44	*Cylindrospermum* sp. *Gomphosphaeria* sp. *Lyngbya* (2 strains) *Microcystis* (6 strains) *Nostoc* (7 strains) *Oscillatoria* (2 strains) *Phormidium* (3 strains) *Plectonema* (7 strains) *Synechococcus cedrorum* *Nostoc muscorum* *Plectonema boryanum*	U.S.A.	Stewart & Brown (1969, 1971)
Actinomycete			
D5, BB-49 BB-53	*Anacystis nidulans* *Anacystis* sp. *Fremyella diplosiphon* *Lyngbya* spp. *Nostoc commune* *N. muscorum* *Nostoc* sp. *Phormidium* sp. *Plectonema boryanum*	U.S.A.	Safferman & Morris (1962, 1963b)
Various			
Streptomyces spp.	Various species	Russia	Rubenchyk *et al.* (1965) Bershova *et al.* (1968)
Fungus			
Blastocladiella anabaenae	*Anabaena flos-aquae* *A. circinalis* *A. solitaria* *Aphanizomenon flos-aquae*	England England England England	Canter & Willoughby (1964) Canter & Willoughby (1964) Canter (1968) Canter (1968)

Table 1—*continued*

Pathogen	Host range	Country of origin	Reference
Fungus			
Chytridium cornutum	*Anabaena circinalis*	Germany	Braun (1856)
C. microcystidis	*Aphanizomenon flos-aquae*	Czechoslovakia	Canter (1963)
	Microcystis aeruginosa	England	Canter (1972)
	M. aeruginosa	Poland	Szklarczyk (1956)
	M. flos-aquae	Scotland	Canter (1972)
	M. aeruginosa	Sweden	Canter (1972)
	M. flos-aquae		
	M. viridis		
Phlyctidium globosum	*Aphanizomenon flos-aquae*	Sweden	Skuja (1956)
P. megastomum	*Anabaena flos-aquae*	Russia	Raitschenko (1902)
Rhizophydium deformans	*Oscillatoria rubescens*	Switzerland	Jaag & Nipkow (1951)
R. megarrhizum	*Aphanizomenon flos-aquae*	U.S.A.	Paterson (1958)
	Lyngbya sp.	U.S.A.	Paterson (1958)
	Oscillatoria agardhii	England	Canter & Lund (1951)
			Canter (1972)
	Oscillatoria sp.	Belgium	DeWildeman (1890)
	Oscillatoria sp.	England	Canter & Lund (1951)
			Canter (1972)
R. subangulosum	*Aphanizomenon gracile*	Czechoslovakia	Fott (1951)
		Denmark	Fjerdingstad (1955)
R. ubiquetum	*Anabaena solitaria*	England	Canter (1968)
R. oscillatoriae-rubescentis	*Oscillatoria rubescens*	Switzerland	Jaag & Nipkow (1951)
Rhizosiphon akinetum	*Anabaena macrospora*	Czechoslovakia	Canter (1954)
	A. affanis	England	Canter (1954)
R. anabaenae	*A. affanis*	Eire	Canter (1953)
	A. circinalis	England	Canter (1972)
	A. macrospora	Czechoslovakia	Fott (1951)
		Germany	Seligo (1909), Canter (1968)
		Sweden	Skuja (1948), Canter (1951)
	A. planktonica	U.S.A.	Paterson (1958)

Table 1—*continued*

Pathogen	Host range	Country of origin	Reference
	A. sphaerica	Sweden	Skuja (1948), Canter (1951)
	A. spiroides	Eire	Canter (1953)
R. crassum	*A. affanis*	Sweden	Skuja (1948), Canter (1951)
		Wales	Canter (1953)
		Eire	Canter (1953)
	A. circinalis	England	Canter (1951)
		Eire	Canter (1953)
	A. solitaria	England	Canter (1953)
	A. spiroides	England	Canter (1968)
		Eire	Canter (1953)
		Northern Ireland	Canter (1953)
	Anabaena spp.	England	Canter (1951, 1953)
	Anabaena sp.	Sweden	Canter (1951)
Rhizophydium sp.	*Oscillatoria* sp.	U.S.A.	Sparrow *et al.* (1965)
Scherffeliomyces sp.	*Anabaena* sp.	U.S.A.	Paterson (1958)
Unknown	*Aphanizomenon flos-aquae*	Scotland	Canter (1972)
	Gomphosphaeria naegeliana	Germany	Seligo (1909)
		England	Canter (1954)
Unnamed Chytrid	*Oscillatoria* sp.	U.S.A.	Paterson (1958)

are seen as polyhedrons 88 nm diam. with 5 or 6 knob-like projections. There is no tail, but an almost indiscernible collar is present. The nucleic acid is double stranded. The pH and thermal stability ranges are very similar to those of other cyanophages (i.e. stable within the pH range 5–11 and up to 40°). Unlike LPP-1, it is also stable in distilled water (i.e. it does not have a Mg^{2+} requirement). The plaques produced by SM-1 are usually < 1 mm diam., but plaque size may vary depending on the algal culture medium used.

Virus AS-1 (Fig. 1) isolated by Safferman *et al.* (1972) infects only *Anacystis nidulans* and *Synechococcus cedrorum* and is inactive against other blue-green

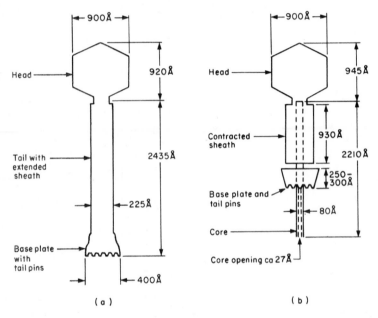

Fig. 1. Structure of cyanophage AS-1. (a) shows the normal appearance of the virus while (b) shows the particle with contracted tail sheath and base plate (from Safferman *et al.*, 1972).

algae, bacteria and actinomycetes. Highest titres are obtained using *Synechococcus cedrorum* as host. Under the electron microscope it appears as a polyhedron (90 nm diam.) with a long rigid tail (243 nm long and 22.5 nm diam.). Short 'pins' are attached to the base of the contractile tail and these may be important in attaching the particle to the host, as in the T2 coliphage. Thermal denaturation tests suggest double stranded DNA, and the G+C ratio is 53– 54% (cf. 55% in LPP-1, Goldstein & Bendet, 1967). The virus is serologically distinct from the other blue-green algal viruses including SM-1 which also infects certain unicellular algae. It also resembles, and is perhaps identical to an isolate

(AS-2) which we have obtained recently from a sewage works in Scotland and which infects *Anacystis nidulans, Synechococcus cedrorum* and *Synechococcus* sp. NRC-1 (Plate 1). It is also more pH-stable than other algal viruses and is temperature-stable at 45°, although infectivity decreases by 33% at 50° and virtually disappears at 60°.

The N-1 virus isolated by Adolph & Haselkorn (1971) is the only virus effective against heterocystous algae which has been characterized in detail. It infects only certain strains of *Nostoc muscorum* and on lawns produces plaques *c.* 3 mm diam. in 3 days. The virus has a hexagonal head (61 nm diam.) and a long, contractile tail (110 nm long and 16 nm wide) which is striated with 9–11 bands/tail. The virus is composed of at least 19 structural proteins, the major of which have molecular weights of 37 000 and 14 000 daltons. The G+C ratio is 37% based on buoyant density measurements and 41% based on melting point temperature determinations. The double stranded DNA has a MW of $44\pm3 \times 10^6$ daltons.

In addition to these cyanophages various others have been reported on briefly. These include particles attacking *Anabaenopsis raciborskii,* certain strains of *Cylindrospermum, Raphidiopsis indica,* (Singh & Singh, 1967; Singh, Singh & Varanasi, 1969), *Anabaena variabilis* (Granhall & Von Hofsten, 1969; Koz'yakov, Gromov & Khudyakov, 1972), *Aphanizomenon flos-aquae* (Granhall, 1972), the occurrence of virus-like particles in *Oscillatoria princeps* (Ueda, 1965), and viruses in Russian freshwaters (Goryushin & Chaplinskaya, 1966, 1968; Rubenchyk *et al.,* 1966). Perhaps the most interesting of these is the supposed virus of *Oscillatoria princeps.* This alga contains rod-shaped structures each *c.* 300 nm long x 19 nm diam. Although Ueda (1965) considered it to be rather similar to tobacco mosaic virus, it is perhaps best to leave it, until it is isolated and characterized, in the same category of unknowns as the variety of virus-like particles described in natural populations of blue-green algae by Jensen & Bowen (1970).

Although this section is concerned mainly with viruses of blue-green algae, one cannot fail to mention that over the last few years more and more information has accumulated on what appear to be viruses of eukaryotic algae. As early as 1961 Zavarzina & Protsenko suggested that they had discovered what may have been viruses of the green alga *Chlorella pyrenoidosa.* These were spherical particles, 30–47 nm diam., and stable at 60° for 10 min. This initial report and a subsequent one (Zavarzina, 1962) was treated with some scepticism at the time, but since then other viruses, or more strictly 'virus-like particles', have been reported in green algae (Pickett-Heaps, 1972), red algae (Lee, 1971) and brown algae (Toth & Wilce, 1972).

In the green alga, *Oedogonium,* the virus can be seen clearly under the electron microscope as a hexagon *c.* 240 nm diam. It has a complex arrangement of the sub-particles and is associated with areas rich in ribosomes (Pickett-Heaps,

1972). The particles reported by Lee (1971) in sections of the red alga *Sirodotia* are polygonal in shape, 50–60 nm diam., while in 3-days old sporelings of the brown alga *Chorda tomentosa* virus-like particles, 170 nm diam., can be seen clearly and these seem to lyse the *Chorda* cells (Toth & Wilce, 1972). Like those reported from other eukaryotic algae, they have not been isolated so far. Viruses of eukaryotic algae present a challenging field which merits much further research.

(b) *Actinomycetes*

Actinomycetes which lyse blue-green algae have been reported from the United States (Safferman & Morris, 1962, 1963*b*) and from Russia (Rubenchyk, Bershova & Knizhnik, 1965; Bershova, Kopteva & Tantsyurenko, 1968). In their early screening tests, Safferman & Morris (1962) found that over 200 actinomycetes which they isolated caused lysis of blue-green algae and, in some cases at least, the mode of action appeared to be via extracellular products (Safferman & Morris, 1963*b*). Rubenchyk *et al.* (1965) found that 28 strains of *Streptomyces* were effective against a bloom-forming *Anabaena* species. It is remarkable that these actinomycetes effective against blue-green algae have not been investigated more thoroughly.

(c) *Bacteria*

In the early 1960s much attention was paid to the possible importance of viruses of blue-green algae in natural ecosystems and it was postulated (Safferman & Morris, 1964*b*) that such viruses may play a major role in regulating the development of algal blooms. However with time it has become apparent that viruses of bloom-forming algae such as *Microcystis, Gloeotrichia* and *Anabaena* seem not to be very widespread (or else we have not developed suitable techniques for detecting them) and that much more frequently the plaques caused on algal lawns are due to algal-lysing bacteria rather than to viruses. Indeed it seems from our studies in the United Kingdom (Daft, McCord & Stewart, 1975) and those of other workers elsewhere (Shilo, 1970; Stewart & Brown, 1971; Gromov *et al.*, 1972) that algal-lysing bacteria are not only abundant in many habitats, but are probably distributed widely throughout the world.

The known algal-lysing bacteria are listed in Table 1. Among the lesser studied types is the *Cellvibrio* investigated in Sweden by Granhall & Berg (1972). This bacterium produces its algal-lysing material only in stationary-phase culture. Similarly, little is known about the importance of the fruiting myxobacteria, such as those studied by Wu, Hamdy & Howe (1968) and Stewart &

Brown (1970), which cause lysis of blue-green algae. In the case of both *Myxococcus* and *Sorangium* the active compound against blue-green algae is liberated extracellularly and has no effect on the green alga *Tetracystis intermedium* (Stewart & Brown, 1970). Another group of bacteria which could be of importance, although we cannot be sure until more data are available, are the flexibacteria of the type reported from Russia by Gromov *et al.* (1972). They studied *Flexibacter flexilis,* an orange-pigmented bacterium with a low G+C ratio (35.9%). This bacterium, which moves by a slow gliding movement, lyses species of *Anabaena, Phormidium* and *Nostoc* and contact with the host is necessary for lysis to occur. Apart from its low G+C ratio, this bacterium resembles the next group of organisms in many respects.

The non-fruiting myxobacteria with high G+C ratios (65–70%), in contrast to the other groups of algal-lysing bacteria, have been studied fairly extensively. Over 30 different strains are known (Shilo, 1971; Stewart & Brown, 1969, 1971; Daft & Stewart, 1971; Daft *et al.,* 1975) and many are closely related. For example, we have isolated 16 different strains from various habitats within the United Kingdom and all of these are serologically related, although the antiserum is ineffective against 30 other non-pathogenic bacteria isolated from similar freshwater habitats (Daft & Stewart, unpubl.). The organisms are all Gram negative rods, with rounded ends, averaging 10.6 x 2.4 μm in size and moving with a slow gliding movement. Flagella are absent. Despite their high G+C ratios these bacteria do not produce fruiting bodies or microcysts. On solid medium they form a silky growth over the agar surface, this growth often being pinkish or yellowish in colour due to the presence of carotenoid pigments, absorbing in the range 410–500 nm (a maximum of 440 nm in methanol). These non-fruiting myxobacteria, as will be discussed later, are of 2 sub-groups which differ in their mode of attack, one sub-group requiring a contact mechanism for lysis, the other effecting lysis by means of soluble extracellular enzyme(s). To date, these organisms do not fit into any group either in the 1957 edition of Bergey's manual (Breed, Murray & Smith, 1957) or in the new edition in press (as far as we know) and the setting up of an additional class into which these organisms can be placed is required.

(d) *Fungi*

The fungi which cause lysis of blue-green algae are all members of the Chytridiales with one exception, *Blastocladiella anabaenae,* which is a member of the Blastocladiales (Table 1). Our knowledge of these organisms is based almost entirely on the detailed work of Canter (see Canter, 1972) but so far their growth and physiology have not been studied in any detail because the organisms cannot be cultured easily in the absence of their hosts. Most have complicated life histories which are as yet not entirely understood although much progress has been made (see e.g. the studies of Canter, 1951, on *Rhizosiphon crassum* which infects certain *Anabaena* species).

3. The Isolation and Examination of Algal Pathogens

The techniques required for the accurate examination and characterization of algal pathogens depend, of course, on the types of organisms being investigated. In the case of the fungi, for example, which have not been maintained in pure culture for long periods (although certain species such as *Rhizophidium megarrhizum* can be maintained in the laboratory for several weeks (Lund, 1957)), important advances can be made simply on the basis of critical examination under the light microscope as shown by the excellent work of Canter (see Canter, 1972).

On the other hand, with viruses it is necessary to isolate them and then to characterize them morphologically under the electron· microscope, followed by purification and biochemical characterization. The techniques used in the isolation of viruses have been considered by Stewart & Daft (1976) and need only be mentioned here in passing. The usual technique is to bring water samples back to the laboratory, treat them with chloroform to remove living cells, remove the chloroform and then place known volumes of water on lawns of algae growing in Petri-dishes. Viral plaques appear, usually within 3–7 days, if a virus which causes lysis of the algal host is present (Plate 2), plaques of different size indicating that different viral strains are present. The virus can then be picked from particular plaques and characterized further. There are modifications of the above technique. For example, the water samples may be concentrated before plating using osmotic techniques (Padan, Shilo & Kislev, 1967), or else potentially susceptible algae can be added to the test water before plating to increase the titre of infective agents. Another technique which has been used is to float pieces of susceptible algae for 1–3 days in the natural habitats where viruses may occur and then to bring these algae back to the laboratory for subsequent examination for pathogens (Daft, Begg & Stewart, 1970). In all cases subsequent characterization of the viruses from the plaques follows the same procedure, and 3 essential criteria must be satisfied before it can be accepted that a lytic cyanophage is in fact present, thus: (i) the infective particle should pass through a bacteria-proof filter and will probably survive chloroform treatment; (ii) the infective particle must be distinguishable as a virus under the electron microscope; (iii) it must be possible to re-infect the host with it, and after subsequent lysis of the host, to re-isolate the particle.

The isolation of lytic bacteria and actinomycetes from natural populations follows exactly the same procedures as those used with viruses except that the organisms are non-filterable and the chloroform treatment step is not used. The bacteria are isolated from the plaques which they produce in the usual way and are further characterized using routine bacteriological techniques.

4. The Ecology of Algal Lysing Agents

Blue-green algae are characteristic of mesotrophic and eutrophic waters with a neutral or alkaline pH and are seldom abundant in acid waters. In general, the occurrence of algal lysing viruses and bacteria parallels the occurrence of the host and both respond similarly to environmental change. Thus for *Microcystis* NRC-1 the optimum pH range is 8–11 (Zehnder & Gorham, 1960), and growth increases up to temperatures of 35°. The cyanophage to it, SM-1, shows a pH range of 5–11, and is temperature stable up to 46°. With *Plectonema boryanum* the pH range is 7–11 and the temperature range is up to 45°, while the corresponding values for its cyanophage LPP-1, are 7–11 and up to 40°, respectively (Safferman & Morris, 1964*a*).

In the case of algal lysing bacteria which are less specific in their host range the response to environmental conditions, although correlating directly with the response of the hosts, is not so tightly coupled, as befits a non-obligate association. For example Daft *et al.* (1975) studied the effect of pH on 5 bacterial isolates (CP-8, CP-10, CP-13, CP-15 and CP-16) and although there was some variability from isolate to isolate the pH optimum for each species was within the range 7–9, that is at levels which are optimum for many planktonic blue-green algae. Nevertheless, as the data in Fig. 2 show, the bacteria grow and cause lysis most rapidly at pH 8.4 although algal metabolism (measured as nitrogenase activity) is optimum at 7.6. Such data also confirm that these myxobacteria lyse actively growing cells and not unhealthy cells only.

One environmental condition which appears to affect the bacteria and the algae quite differently is the response to oxygen, because the bacteria are obligate aerobes while Cyanophyceae in general are inhibited by high oxygen concentrations which stimulate photorespiration (Lex, Silvester & Stewart, 1972) and which may also lead to photo-oxidative death (Abeliovich & Shilo, 1972). In view of the close correlation of cyanophage LPP-1, its host and the eutrophic state of the water Shane, Cannon & De Michele (1972) have suggested that plaque counts of the LPP-virus may be a sensitive bioassay method for determining the eutrophic state of the environment. However, we question whether this is a feasible proposition because the eutrophic state of the water is not the only factor which regulates algal growths; other physical, chemical and biological parameters are all important. Also as various workers have reported (Safferman & Morris, 1967; Padan & Shilo, 1969; Daft *et al.*, 1970), viral numbers fluctuate greatly in different parts of even the same body of water and there may also be diurnal fluctuations in viral titres (MacKenzie & Haselkorn, 1972). There is also the complication of virulent and temperate phages (see Singh *et al.*, 1969; Cannon & Shane, 1972; Cannon, Shane & Bush, 1971; Padan, Shilo & Oppenheim, 1972) with some viruses being lytic, for example, in the field but not in the laboratory and vice versa, and on one host but not on

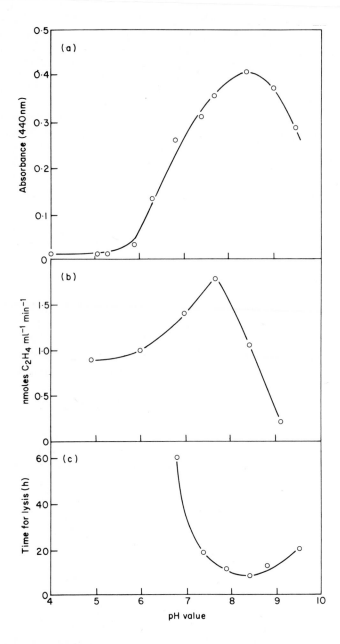

Fig. 2. The effect of pH on: (a) the growth of lytic bacterium CP-1, (b) nitrogenase activity of the host, *Nostoc ellipsosporum,* and (c) the rate of lysis of the host.

another. In this section we will consider the various habitats where algal lysing agents have been investigated in most detail.

(a) *Waste stabilization ponds*

These are probably the areas where cyanophages occur in highest numbers, both in terms of total titre and numbers of different types recorded. For example the original LPP-1 virus was isolated from a waste stabilization pond in south eastern Indiana (Safferman & Morris, 1963*a*), the morphologically similar but serologically distinct LPP-2 also came from a sewage pond in Indiana (Safferman *et al.*, 1969*a*) and our D-1 isolate was obtained from a sprinkler type of sewage works at Forfar in Scotland (Daft *et al.*, 1970). The occurrence of the LPP-viruses in waste stabilization ponds in the United States was also noted by Jackson (1967) and Jackson & Sladecek (1970) while Shane (1971) in a survey of part of the Delaware river found highest titres in sewage oxidation ponds. Cyanophages also occur in sewage oxidation ponds in India. In addition to the LPP types, agents which lyse heterocystous species of *Anabaena* and *Cylindrospermum* have been noted (Singh & Singh, 1967). The viruses of unicellular algae, SM-1 and AS-1, came respectively from sewage oxidation ponds in Indiana (Safferman *et al.*, 1969*b*) and Florida (Safferman *et al.*, 1972).

The general abundance of LPP viruses in sewage works can be gauged from the studies of Safferman & Morris (1967) who detected the particles in 11 of 12 ponds studied in the United States. There the viruses showed a distinct seasonal periodicity with numbers fluctuating during the year from 4–270 ml^{-1}. A similar type of seasonal variation occurs in the case of D-1 and AS-2 viruses in Scotland (Daft & Stewart, unpublished).

Algal lysing bacteria have also been isolated from sewage works. For example Daft & Stewart (1971) obtained their original isolate CP-1 from a sewage works in Scotland. The abundance of these bacteria varies depending both on the season, and on the area of the sewage works sampled, as shown by the data in Table 2.

(b) *Lakes and reservoirs*

Algal lysing viruses, bacteria and fungi have all been isolated from freshwater lakes and reservoirs. Among the cyanophages, N-1 was isolated from along the shores of Lake Mendota in Wisconsin (Adolph & Haselkorn, 1971). There it occurred round the lake edge and showed a distinct seasonal periodicity with maximum numbers occurring in May. *Nostoc muscorum*, its host, is not a planktonic species in Lake Mendota (or elsewhere) but occurs around the lake edge as noted by one of us (W.D.P.S.). Other heterocystous algae which are abundant in Lake Mendota e.g. *Gloeotrichia, Anabaena* and *Aphanizonenon*

Table 2

*Concentrations of algal lysing bacteria found at different sites at
Forfar sewage works*

Sampling sites	Number of lytic bacteria (p.f.u.* ml^{-1})	
	Autumn (27/8/72)	Winter (15/2/73)
Crude sewage	112	1
Effluent from filter bed	1290	2
Humus tank	1010	2
Final effluent	692	6
Stream water and effluent	307	18
Forfar Loch	267	0

* p.f.u. = plaque-forming units.
(Daft *et al.* 1975).

(Stewart, Fitzgerald & Burris, 1967) are unaffected by this virus, but in the eutrophic Lake Erken in Sweden, *Aphanizomenon flos-aquae* appears to be susceptible to virus attack although the characteristics of the virus have not been described (Granhall, 1972). Viruses attacking species of the unicellular *Microcystis* have been reported from reservoirs in Russia by Goryushin & Chaplinskaya (1966, 1968). From the details which are available these viruses appear to be very similar to the SM-1 virus of Safferman *et al.* (1969*b*). LPP-1

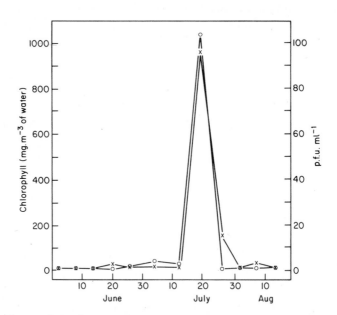

Fig. 3. The correlation between the levels of chlorophyll *a* (o – o) and lytic bacteria (x – x) in the surface waters of Thriepley Loch, Scotland, during the summer of 1972.

viruses also occur in Lake Kinneret in Israel (Padan & Shilo, 1969). Singh (1973) has reported briefly on the occurrence of the LPP virus in various ponds in India. The latter plaque-forming units were detected 3 days after plating and numbers were highest in samples of water collected in March.

Algal lysing bacteria have been isolated from English lakes, Welsh reservoirs and from Scottish lochs (Daft & Stewart, 1971; Daft *et al.*, 1975). In spot tests on Windermere in England the bacteria were associated with a growth of *Oscillatoria,* and in Esthwaite Water and Loweswater with species of *Anabaena, Gomphosphaeria* and *Microcystis* (Daft *et al.,* 1975). In our studies in Scotland we have examined 25 lochs for algal lysing bacteria and have found these in all the lochs where blue-green algae were common (14 of 25). All the bacteria isolated were similar, morphologically and physiologically, to other non-fruiting

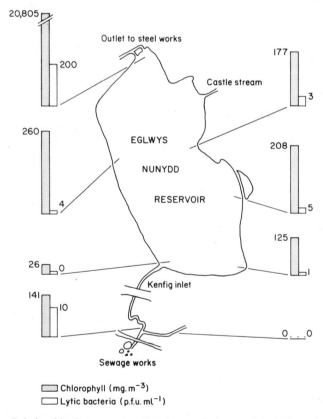

Fig. 4. Relationship between the abundance of algae and lytic bacteria in samples collected from the surface 10 cm of the water column (except for the outlet sample which was surface scum) during a 2 h period at Eglwys Nunydd Reservoir in South Wales in the summer of 1971.

myxobacteria with high G+C ratios which we had already isolated (Daft & Stewart, 1971; Daft *et al.*, 1975). The abundance of the bacteria in these habitats correlates directly with the concentrations of algae in the water samples and such a correlation is clear from our data for Thriepley Loch in Scotland (Fig. 3), and from data for Eglwys Nunydd reservoir in Wales (Fig. 4). Treated sewage provides the main inoculum of lytic bacteria to the latter reservoir.

The correlation between seasonal variation in algal numbers and lytic bacteria is shown by our data for Monikie reservoir in Scotland (Daft, McCord & Stewart, 1973). Here samples were taken from the open reservoir, the reservoir filters and from the filter medium. It was found that the abundance of the algae fluctuated markedly (*Gloeotrichia* predominated in July followed by *Microcystis, Aphanizomenon* and *Anabaena,* until algal growth ceased in November) and the numbers of lytic bacteria changed in a similar manner. The highest concentrations of lytic bacteria occurred in the filter medium after draining $(59\ 000\ \text{ml}^{-1})$ followed by the filters $(2400\ \text{ml}^{-1})$ and finally by water from the open reservoir $(92\ \text{ml}^{-1})$.

Fungal pathogens of various Cyanophyceae have been isolated from various bodies of water in the English Lake district including Esthwaite Water, Windermere (North and South basins), Crossmere, Blake Mere, Kettle Mere, Whitemere, Hatchmere, Loweswater, Blelham Tarn and Loughrigg Tarn (see Canter, 1972 for references). The point here is that they are probably extremely widespread in distribution, and the numerous records of their abundance in the English lakes are available simply because an expert who can identify these fungi has examined these particular waters. There are also reports of their occurrence in Germany, Czechoslovakia, Sweden, the United States, Wales, Scotland, Denmark, Poland, Switzerland, Belgium, Hungary, Ireland, and Russia (see Canter, 1972 for references). The abundance of these fungal pathogens correlates directly with the abundance of their hosts and it does seem that the fungi may cause, in part at least, the death of the hosts on which they depend.

(c) *Fish ponds*

Studies in Israel have shown that fish ponds are a rich source of algal lysing agents. Padan *et al.* (1967) isolated virus LPP-1G, which is identical in shape, size and host range to LPP-1, from such a habitat, although the numbers were lower than those recorded in American waste stabilization ponds. Padan & Shilo (1969) suggest that these cyanophages may be important in regulating the periodicity of algal blooms, although they found highest numbers of viruses when the algae were also present in greatest abundance. The occurrence of these viruses in fish ponds of high salinity makes it probable that they will also be present in truly marine environments. Shilo (1970) has isolated and studied in detail myxobacter FP-1 which was found in waterblooms of green algae in fish

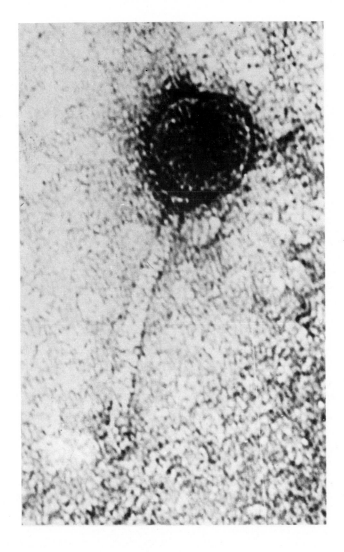

Plate 1. Electron micrograph of cyanophage AS-2. The head is 50 nm diam., the tail length is 130 nm.

[*facing p. 90*

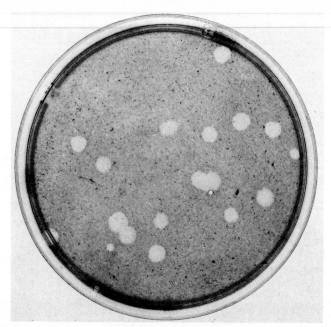

Plate 2. Plaques produced by an LPP type virus obtained from the River Ganges (original material provided by Dr. H. N. Singh). × 1.

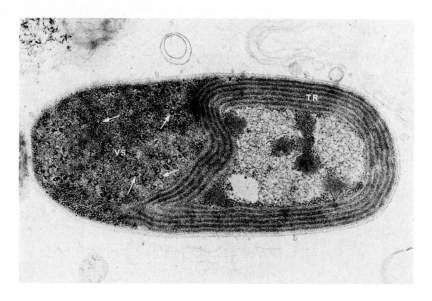

Plate 3. A cell of the blue-green alga *Plectonema boryanum* infected with LPP isolate, D-1. The virogenic stroma (VS) with virus particles (arrows) occurs outside the thylakoidal region (TR). × 22 500.

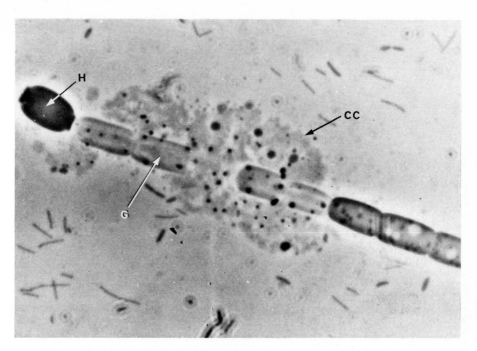

Plate 4. Filament of *Aphanizomenon flos-aquae* attacked by bacterium CP-1 showing extruded cell contents (CC), ghosts of lysed vegetative cells (G) and heterocyst (H). x 3000.

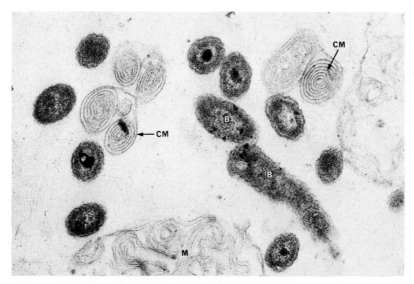

Plate 5. Cellular remains of *Nostoc ellipsosporum* after lysis by bacterium CP-1 (B). Note that only membranes (M) (some of which are coiled, CM) are abundant. x 24 000.

ponds in the Jezrael Valley of Israel. Studies on this organism have provided interesting data on lytic myxobacteria.

(d) *Rivers*

Shane (1971) has studied the distribution of LPP viruses in several Delaware rivers (the Christina, Red Clay and White Clay rivers). Cyanophages were absent from the head waters but they increased in quantity as the rivers flowed through urban areas. They were also found in household ponds, farms, quarries, waste stabilization lagoons etc. and appeared to be resistant to extreme environmental parameters including 70 mg l^{-1} of chloride. The detection of these viruses in rapidly moving waters indicates that they must be multiplying rapidly in such situations.

(e) *Soils*

It would be quite erroneous if we gave the impression that algal lysing agents are found only in aquatic habitats. That is not so; they occur anywhere that their hosts grow, including soils. Thus Jackson & Sladecek (1970) obtained some evidence that the LPP virus can exist in soil and, in India, Singh (1973) found up to 1000 LPP particles ml^{-1} in rice paddy soils. In Scotland we have isolated algal lysing bacteria of the CP-type from various agricultural soils and sand dunes, in fact from virtually any habitat of neutral or alkaline pH. As most of these soils do not contain large numbers of blue-green algae it is probable that the lytic bacteria also lyse other organisms or obtain suitable nutrients elsewhere.

5. The Mode of Action of Algal Lysing Agents

The mode of action of cyanophages on their hosts has been investigated most exhaustively using virus LPP-1. The pattern of growth with its one step growth curve is similar irrespective of whether the numbers of extracellular or intracellular particles are measured (Padan, Ginzburg & Shilo, 1970; Sherman & Haselkorn, 1971). This shows a latent period, a rise period, and finally a viral burst. The burst size varies usually between 100–400 viruses/cell.

The first stage observed under the electron microscope is the attachment of the LPP-1 virus by its tail to the host wall and in size the attached virus tail is longer than that observed in free-living particles (Smith *et al.*, 1966b; Padan *et al.*, 1967). After injection of viral DNA into the host, empty viral heads or ghosts can be seen still attached to the host. On infection the first visible sign of change in the host cell of *Plectonema,* under both the light and electron microscopes is the invagination of the photosynthetic membranes (usually within 3 h) with the virogenic stroma developing outside these invaginated thylakoids (see Plate 3). The synthesis and assembly of the virus within *Plectonema* is complex and there are conflicting views on the pattern which is followed (Brown, Smith & Walne 1966; Smith *et al.,* 1966*a*; Smith, Brown & Walne, 1967; Sherman & Haselkorn, 1970). According to Brown *et al.* (1966) and Smith *et al.* (1967) viral DNA is synthesized in the nucleoplasm of the host. This then moves to the thylakoid region where helices (18 nm diam.) develop and then to the outside of the photosynthetic lamellae (the virogenic stroma) where the protein coat is synthesized. The cyanophage tails are produced just before lysis. Sherman & Haselkorn (1970), on the other hand, believe that the viruses develop entirely in the virogenic stroma in *Plectonema.* When lysis occurs it is usually within 7–10 h of infection. There is also some evidence that the intracellular virus may inject its DNA into an adjacent cell before lysis of the first infected cell (Smith *et al.,* 1966*b*).

The pattern of infection of *Synechococcus* by the SM-1 virus is quite distinct from that of LPP-1 on *Plectonema* in that no invagination of the photosynthetic lamellae occurs and the viruses develop entirely in the nucleoplasm rather than external to the photosynthetic lamellae (Padan *et al.,* 1967). This may explain why infection by the SM-1 virus has no specific effect on photosynthetic CO_2 fixation in infected cells of the host (MacKenzie & Haselkorn, 1972).

The physiological features of the host which regulate virus multiplication have been studied in detail (Padan & Shilo, 1973). Photosynthetic processes are involved in the case of LPP-1 with the burst size in the dark being only *c.* 8% of that in the light. The main role of light appears to be to supply ATP and this can be generated entirely by cyclic photophosphorylation because there is no reduction in burst size in the presence of the inhibitor of photosystem II, DCMU (Padan *et al.,* 1970). Unlike the LPP virus, both N-1 and SM-1 are extremely sensitive to DCMU, indicating a requirement for photosystem II, probably for products of CO_2 fixation for replication (Sherman & Haselkorn, 1971). It seems to us that the different effects of DCMU may be attributable in part to different pre-treatments of the algae prior to the addition of DCMU. For example the addition of DCMU to carbon-depleted cells is likely to have more effect on metabolism than is its addition to carbon-rich cells (Lex & Stewart, 1973) and detailed comparative studies on the role of photosystem II on the LPP, SM-1, and N-1 viruses are required.

The bacteria which lyse blue-green algae do so by 2 main mechanisms. Bacteria such as CP-1 and FP-1 require contact with the host for lysis to occur while others such as *Cellvibrio, Flexibacter flexilis, Myxococcus xanthus* and *Myxobacter* 44 lyse the algae by producing a soluble extracellular compound(s).

The events leading to lysis by FP-1 (Shilo, 1970) and CP-1 (Daft & Stewart, 1971, 1973) are very similar and can be considered together. For lysis to occur the bacteria must be metabolically active and for this they require

O_2, being strict aerobes. When added to cultures of algae they move towards the photosynthetically O_2-evolving algae and attach themselves end-on to the algal cell wall. In filamentous algae attachment is often, but not always, close to the cross septae and with heterocystous algae such as *Anabaena* the bacteria often aggregate round the heterocyst which, however, is not lysed rapidly. Sometimes the bacteria become attached end-on to each other and time-lapse photography shows that this end-on attachment of bacteria to each other is not due to division of attached bacteria. There may be some factor regulating the number of bacteria which become attached to the vegetative cells, because after several become attached to a particular cell few others do so, even although cell wall surface is still available, and they tend to move on to a less populated cell.

Within 30 min of attachment under optimum environmental conditions changes are seen in the appearance of the host. In *Aphanizomenon flos-aquae* the algal protoplast becomes paler in colour, while in *Oscillatoria redekei* the highly refractive appearance of the cells is lost and gas vacuoles disappear. In the case of *Anabaena circinalis* the protoplasmic contents including the gas vacuoles are lost from the cell within an hour of attachment and often all that remains of the host cell are outlines or 'ghosts' of lysed cells (Plate 4).

Electron microscope studies show that one of the earliest stages of lysis is the loss of the electron dense L_2 layer of the cell wall. The L_4 layer remains initially but this together with the L_1 and L_3 layers eventually disappears leaving only spherical protoplasts. The plasmalemma then ruptures either by random breakage or, as sometimes happens, for example in *Nostoc ellipsosporum,* the plasmalemma coils inwards around itself (Plate 5) and the cell contents are then released into the medium leaving mainly membranous remains and lipid droplets. The pattern of lysis in other algae may differ slightly (Daft & Stewart, 1973).

The bacteria produce a variety of enzymes, some or all of which may be important in the lytic process. The rapid dissolution of the L_2 layer of the wall indicates that a lysozyme type of enzyme is important and the organisms show proteolytic activity as evidenced by their capacity to liquify gelatine. They can also break down peptone, casitone and soluble starch (Daft *et al.,* 1975). A contact mechanism may be particularly advantageous to these bacteria because it results in a concentration of the enzymes at the point where they are required most. In this way the enzymes are not diluted out as would happen with an extracellularly liberated enzyme in an aquatic system. Certain bacteria depend, however, on truly extracellular enzymes for lysis and the most detailed study on this type of organism has been carried out by Stewart & Brown (1971) on *Myxobacter* 44. This bacterium which produces a blackish-brown pigment releases an extracellular lysozyme-like enzyme and bacteria-free filtrates cause lysis. In addition to its lysozyme activity *Myxobacter* 44 shows proteolytic, alginase, cellulase and chitinase activities (Stewart & Brown, 1971).

6. The Possible Use of Lytic Agents in the Control of Algal Growths

Blue-green algae are nuisance algae in many mesotrophic and eutrophic waters. They can cause deoxygenation of the waters, thereby killing fish, they may cause offensive odours and affect the amenity value of the waters, they are not eaten readily by most animals and thus accumulate. They are also particularly resistant to chemical algicides such as copper sulphate, possibly because the extracellular materials which they produce form complexes with the copper (Fogg *et al.*, 1973). Methods for their control are thus a matter of some concern and the use of biological agents to prevent or destroy growths of blue-green algae on a routine basis is an intriguing, if generally unproven, possibility.

The requirements for a satisfactory algicide and the extent to which cyanophages, lytic bacteria, and copper sulphate meet these are listed in Table 3. It is seen that of the criteria listed cyanophages appear to offer in theory a much better prospect for the control of blue-green algae than do the bacteria, although the latter also have some advantages over the use of copper sulphate. It is

Table 3

The potential merits of cyanophages and lytic bacteria as algicidal agents compared with copper sulphate

Properties	Cyanophages	Lytic bacteria	Copper sulphate
Efficient	+++++	++++	+++
Highly selective	+++++	+++	+
Non-toxic to other organisms	+++++	+++++	+
Chemically inert	+++++	++++	+
Cheap	+++++	+++	++
Easy to maintain and apply	+++++	+++	+
Readily available	++	++	++++
Easy to monitor	+++++	+++++	+++++

The degree of efficiency or suitability is indicated by the number of '+' signs.

determining whether these agents are satisfactory in practice which has stimulated much work in this area, and it is worth considering the data available to date.

Safferman & Morris (1964*b*) were the first to carry out mass inoculation studies when they added the LPP-1 virus to simulated natural blooms of *Plectonema* which they had arranged in 112 x 1 l batches. They found that after a week the growth of *Plectonema* had disappeared completely and that the virus titre in the water had increased 3000-fold. Jackson & Sladecek (1970), also using a model system, obtained some evidence that LPP viruses effectively controlled the growth of *Plectonema* in 5000-gallon tanks at a sewage works in New York. They found that they could grow *Plectonema* successfully in autoclaved water, but despite over 40 attempts they could not get it to grow in non-sterilized medium. They therefore concluded that the LPP-1 virus occurred normally in these tanks and that this infected and caused the death of added *Plectonema.* It may be noted that in the experiments of both Safferman & Morris (1964*b*) and Jackson & Sladecek (1970) excess nutrients were available to the algae so that lack of them was not due to nutrient deprivation.

The possible use of cyanophages as a means of algal control should not be accepted non-critically because algal strains resistant to viral attack have been reported to occur at frequencies of *c.* 1 in 10^7 (Padan *et al.,* 1967; Gromov & Koz'yakov, 1970). Resistance is thought to be brought about by changes in the cell envelope which prevent attachment of the cyanophages (Padan *et al.,* 1967). Nevertheless it does seem that the viruses too can overcome this problem as evidenced by the fact that in the field *Plectonema,* for example, is usually found in low numbers, and there is evidence from various laboratory studies and those of Jackson & Sladecek (1970) that cyanophages can effectively reduce the *Plectonema* populations. What has not been done yet is to continue such tests for very long periods to determine whether, in large scale tests, resistant strains develop which then grow profusely to the original algal level. Such tests are essential towards a proper understanding of the usefulness of cyanophages in bloom control.

What may be a more important problem than algal mutation is the finding that only LPP viruses are common. Nevertheless it is this very point which makes their potential use against bloom-forming species attractive. If viruses against the latter can be isolated more readily than at present, and are virulent, then their introduction into bodies of water could be very effective in controlling bloom formation. Indeed there are reports from Russia that the addition of virus has been effective in clearing a reservoir of *Microcystis* (Sirenko, 1973).

The use of algal lysing bacteria instead of virus would overcome the problem of host specificity and indeed we have successfully used our CP-1 bacterium to clear a *Microcystis* bloom from waters (200 l) which we isolated at the edge of a reservoir (Eglwys Nunydd) in Wales. The bacterium was grown in mass culture in the laboratory and was added at a final concentration of bacterial cells of 10^6 ml^{-1}, to one isolated area but only sterile bacterial medium was added to an adjacent control area. It was found that within 3 days complete lysis of the algae to which the bacteria had been added occurred, while in the control series the algae remained healthy. Thus the lysis of bloom-forming blue-green algae on adding lytic bacteria is feasible both in the laboratory and in the field. The problem is that the amounts of bacteria which we had to add were many times higher than the highest concentrations which we find in lakes and reservoirs in the field. Furthermore the bacteria do not depend on lysis of whole algae for their growth and thus when other organic compounds become available their effectivity drops off.

We would emphasize that in field studies strict controls are absolutely essential, because algal blooms are unpredictable in their appearance and disappearance, even within a matter of hours. Thus it is not sufficient simply to add, for example, a cyanophage to a single pond or reservoir and to attribute any subsequent disappearance of algae to cyanophage attack. If lysis does not occur it can probably be said that the cyanophage is ineffective, but if the bloom disappears it can be attributed to cyanophages only if the control series run at the same time is not lysed. Algal blooms are a fascinating but complex subject to which exacting science must be applied. Thus it is essential to accumulate basic data on the taxonomy, ecology, physiology and biochemistry of algal lysing agents and their hosts today if it is hoped that biological control mechanisms can be used to regulate the algal growths of tomorrow.

7. Acknowledgements

Our own work discussed here was made possible through research support from S.R.C., N.E.R.C. and the Royal Society. We thank Miss Susan McCord for technical assistance.

8. References

ABELIOVICH, A. & SHILO, M. (1972). Photo-oxidative death in blue-green algae. *J. Bact.* **111**, 682.

ADOLPH, K. W. & HASELKORN, R. (1971). Isolation and characterization of a virus infecting the blue-green alga *Nostoc muscorum. Virology* **46**, 200.

ADOLPH, K. W. & HASELKORN, R. (1973). Isolation and characterization of a virus infecting a blue-green alga of the genus *Synechococcus. Virology* **54**, 230.

BERSHOVA, O. I., KOPTEVA, Zh.P. & TANTSYURENKO, E. V. (1968). The inter-relations between the blue-green algae – the causative agents of water "bloom" – and bacteria. In *Tsvetenie Vody.* Ed. A. V. Topachevsky (Russian). Kiev, U.S.S.R.: Naukova Dumka.

BRAUN, A. (1856). Über *Chytridium,* eine Gattung einzelliger Schmarotzergewach se auf Algen und Infusorien. *Abhandl. Berlin Akad.* **1855**, 21.

BREED, R. S., MURRAY, E. G. D. & SMITH, N. R. (1957). *Bergey's Manual of Determinative Bacteriology.* Seventh Edition. Edinburgh & London: E. & S. Livingstone Ltd.

BROWN, R. M. (1972). Algal viruses. *Advan. Virus Res.* **17**, 243.

BROWN, R. M., SMITH, K. M. & WALNE, P. L. (1966). Replication cycle of the blue-green algal virus LPP-1. *Nature, Lond.* **212**, 729.

CANNON, R. E. & SHANE, M. S. (1972). The effect of antibiotic stress on protein synthesis in the establishment of lysogeny of *Plectonema boryanum. Virology* **49**, 130.

CANNON, R. E., SHANE, M. S. & BUSH, V. N. (1971). Lysogeny of a blue-green alga *Plectonema boryanum. Virology* **45**, 149.

CANTER, H. M. (1951). Fungal parasites of the phytoplankton. II. Studies on British Chytrids XII. *Ann. Bot.* **15**, 129.

CANTER, H. M. (1953). Annotated list of British aquatic chytrids. *Trans. Br. mycol. Soc.* **36**, 278.

CANTER, H. M. (1954). Fungal parasites of the phytoplankton. III. *Trans. Br. mycol. Soc.* 37, 111.

CANTER, H. M. (1963). Concerning *Chytridium cornutum* Braun. *Trans. Br. mycol. Soc.* 46, 208.

CANTER, H. M. (1968). Studies on British chytrids XXVII. *Rhizophydium fugax* sp. nov., a parasite of planktonic cryptomonads with additional notes and records of planktonic fungi. *Trans. Br. mycol. Soc.* 51, 699.

CANTER, H. M. (1972). A guide to the fungi occurring on planktonic blue-green algae. In *Taxonomy and Biology of Blue-Green Algae.* Ed. T. V. Desikachary. University of Madras.

CANTER, H. M. & LUND, J. W. G. (1951). Studies on plankton parasites. III. Examples of the interaction between parasitism and other factors determining the growth of diatoms. *Ann. Bot.* (N.S.). 15, 359.

CANTER, H. M. & WILLOUGHBY, L. G. (1964). A parasitic *Blastocladiella* from Windermere plankton. *J. R. microsc. Soc.* 83, 365.

DAFT, M. J. & STEWART, W. D. P. (1971). Bacterial pathogens of freshwater blue-green algae. *New Phytol.* 70, 819.

DAFT, M. J. & STEWART, W. D. P. (1973). Light and electron microscope observations on algal lysis by bacterium CP-1. *New Phytol.* 72, 799.

DAFT, M. J., BEGG, J. & STEWART, W. D. P. (1970). A virus of blue-green algae from freshwater habitats in Scotland. *New Phytol.* 69, 1029.

DAFT, M. J., McCORD, S. & STEWART, W. D. P. (1973). The occurrence of blue-green algae and lytic bacteria at a waterworks in Scotland. *Water Treatment and Examination,* 22, 114.

DAFT, M. J., McCORD, S. & STEWART, W. D. P. (1975). Ecological studies on algal lysing bacteria in freshwaters. *Freshwater Biology* (in press).

DE WILDEMAN, E. (1890). Chytridiacées de Belgique. *Ann. Soc. Belge Micro. (Mem.)* 14, 5.

ENSIGN, J. C. (1971). In SHILO, M. (1971), Biological agents which cause lysis of blue-green algae. *Mitt. Internat. Verein. Limnol.* 19, 206.

FJERDINGSTAD, E. (1955). *Rhizophidium deformans* en algeparasit. i *Oscillatoria*-arter fra Møllcēaen. *Bot. Tidsskr.* 52, 169.

FOGG, G. E., STEWART, W. D. P., FAY, P. & WALSBY, A. E. (1973). *The Blue-Green Algae.* London & New York: Academic Press.

FOTT, B. (1951). New chytrids parasitizing on algae. *Mem. Soc. r. Sci. Boheme* 4, 1.

GOLDSTEIN, D. A. & BENDET, I. J. (1967). Physical properties of the DNA from the blue-green algal virus LPP-1. *Virology* 32, 614.

GORYUSHIN, V. A. & CHAPLINSKAYA, S. M. (1966). Existence of viruses of blue-green algae. *Mikrobiol. Zh. Akad. Nauk. Ukr.,* RSR 28, 94 (Ukranian).

GORYUSHIN, V. A. & CHAPLINSKAYA, S. M. (1968). The discovery of viruses lysing blue-green algae in the reservoirs of the River Dnieper. In *Tsvetenie Vody.* Ed. A. V. Topachevsky. (Russian). Kiev, U.S.S.R.: Naukova Dumka.

GRANHALL, U. (1972). *Aphanizomenon flos-aquae:* infection by cyanophages. *Physiol. Pl.* 26, 332.

GRANHALL, U. & BERG, B. (1972). Antimicrobial effects of *Cellvibrio* on blue-green algae. *Arch. Mikrobiol.* 84, 234.

GRANHALL, U. & Von HOFSTEN, A. (1969). The ultrastructure of a cyanophage attack on *Anabaena variabilis. Physiol. Pl.* 22, 713.

GROMOV, B. V. & KOZ'YAKOV, S. (1970). A study of the peculiarities of the interrelationship between a blue-green algal population, *Plectonema boryanum* and cyanophage LPP-1. *Bulletin Leningrad University,* 3, 128.

GROMOV, B. V., IVANOV, O. G., MAMKAEVA, K. A. & AVILOVA, I. A. (1972). A flexibacterium lysing blue-green algae. *Mikrobiologiya* 41, 1074.

JAAG, O. & NIPKOW, F. (1951). Neue and wenig bekannte parasitische pilze auf Planktonorganismen schweizerischer Gewässer I. *Ber. schweiz. bot. Ges.* 61, 478.

JACKSON, D. (1967). Interaction between algal populations and viruses in model pools – a possible control for algal blooms. *ASCE Ann. Nat. Meet. Water Resources Eng.* prelim. draft, 17.

JACKSON, D. & SLADECEK, V. (1970). Algal viruses: eutrophication control potential. *Yale Scientific Magazine* **44**, 16.

JENSEN, T. E. & BOWEN, C. C. (1970). Cytology of blue-green algae. II. Unusual inclusions in the cytoplasm. *Cytologia* **35**, 132.

KOZ'YAKOV, S. Y., GROMOV, B. V. & KHUDYAKOV, I. Y. (1972). Cyanophage A-1(L) of the blue-green alga *Anabaena variabilis. Microbiologiya* **41**, 555.

KRAUSS, R. W. (1960). *Transaction of 1960 Seminar on Algae and Metropolitan Wastes.* Robert A. Taft Sanitary Engineering Centre. Technical Report W61-3. Cincinatti, Ohio: U.S. Public Health Service.

LEE, R. E. (1971). Systemic viral material in the cells of the freshwater red alga *Sirodotia tenuissima* (Holden) Skuja. *J. cell Sci.* **8**, 623.

LEX, M., SILVESTER, W. B. & STEWART, W. D. P. (1972). Photorespiration and nitrogenase activity in the blue-green alga *Anabaena cylindrica. Proc. R. Soc. Lond. B.* **180**, 87.

LEX, M. & STEWART, W. D. P. (1973). Algal nitrogenase, reductant pools and Photosystem 1 activity. *Biochim. biophys. Acta.* **292**, 436.

LUFTIG, R. & HASELKORN, R. (1967). Morphology of a virus of blue-green algae and properties of its deoxyribonucleic acid. *J. Virol.* **1**, 334.

LUND, J. W. G. (1957). Fungal diseases of plankton algae. In *Biological aspects of the Transmission of Disease.* Ed. C. Horton-Smith. Edinburgh & London: Oliver & Boyd.

MacKENZIE, J. J. & HASELKORN, R. (1972). Photosynthesis and the development of blue-green algal virus SM-1. *Virology* **49**, 517.

PADAN, E., GINZBURG, D. & SHILO, M. (1970). The reproductive cycle of cyanophage LPP-1G in *Plectonema boryanum* and its dependence on photosynthetic and respiratory systems. *Virology* **40**, 514.

PADAN, E. & SHILO, M. (1969). Distribution of cyanophages in natural habitats. *Verh. Internat. Verein. Limnol.* **17**, 747.

PADAN, E. & SHILO, M. (1973). Cyanophages – viruses attacking blue-green algae. *Bact. Rev.* **37**, 343.

PADAN, E., SHILO, M. & KISLEV, N. (1967). Isolation of "Cyanophages" from freshwater ponds and their interaction with *Plectonema boryanum. Virology* **32**, 234.

PADAN, E., SHILO, M. & OPPENHEIM, A. B. (1972). Lysogeny of the blue-green alga *Plectonema boryanum* by LPP cyanophage. *Virology* **47**, 525.

PATERSON, R. A. (1958). Parasitic and saprophytic phycomycetes which invade planktonic organisms. II. A new species of *Dangeardia* with notes on other lacustrine fungi. *Mycologia* **50**, 453.

PICKETT-HEAPS, J. D. (1972). A possible virus infection in the green alga *Oedogonium. J. Phycol.* **8**, 44.

RAITSCHENKO, A. A. (1902). Ueber eine Chytridiaceae; *R. sphaerocarpum* (Zopf) Fischer. *Izv. imp. S-Peterb. bot. Sada* **2**, 124.

RUBENCHYK, L. I., BERSHOVA, O. I. & KNIZHNIK, Zh.P. (1965). On the interrelation of *Anabaena* with bacteria and actinomycetes. In *Ecologia i physiologia sinezelenych vodorosleiy.* Moscow: Nauka. (Russian).

RUBENCHYK, L. I., BERSHOVA, O. I., NOVYKOVA, N. S. & KOPTEVA, Zh.P. (1966). Lysis of the blue-green alga *Microcystis pulverea. Mikrobiol. Zh. Akad. Nauk. Ukr.* RSR **28**, 88. (Ukranian).

SAFFERMAN, R. S. & MORRIS, M. E. (1962). Evaluation of natural products for algicidal properties. *Appl. Microbiol.* **10**, 289.

SAFFERMAN, R. S. & MORRIS, M. E. (1963a). Algal virus: isolation. *Science, N.Y.* **140**, 679.

SAFFERMAN, R. S. & MORRIS, M. E. (1963b). The antagonistic effects of Actinomycetes on algae found in Waste Stabilisation Ponds. *Bact. Proc.* p. 14.

SAFFERMAN, R. S. & MORRIS, M. E. (1964a). Growth characteristics of the blue-green algal virus LPP-1. *J. Bact.* **88**, 771.

SAFFERMAN, R. S. & MORRIS, M. E. (1964b). Control of algae with viruses. *J. Am. Water Works Assoc.* **56**, 1217.
SAFFERMAN, R. S. & MORRIS, M. E. (1967). Observations on the occurrence, distribution and seasonal incidence of blue-green algal viruses. *Appl. Mikrobiol.* **15**, 1219.
SAFFERMAN, R. S., MORRIS, M. E., SHERMAN, L. A. & HASELKORN, R. (1969a). Serological and electron microscopic characterization of a new group of blue-green algal viruses (LPP-2). *Virology* **39**, 775.
SAFFERMAN, R. S., SCHNEIDER, I. R., STEERE, R. L., MORRIS, M. E. & DIENER, T. O. (1969b). Phycovirus SM-1: a virus infecting unicellular blue-green algae. *Virology* **37**, 386.
SAFFERMAN, R. S., DIENER, T. O., DESJARDINS, P. R. & MORRIS, M. E. (1972). Isolation and characterization of AS-1, a phycovirus infecting the blue-green algae, *Anacystis nidulans* and *Synechococcus cedrorum. Virology* **47**, 105.
SELIGO, A. (1909). *Tiere und Pflanzen des Seenplanktons.* Stuttgart: Franchkhische Verlag.
SHANE, M. S. (1971). Distribution of blue-green algal viruses in various types of natural waters. *Water Res.* **5**, 711.
SHANE, M. S., CANNON, R. E. & DeMICHELE, E. (1972). Pollution effects on phycovirus and host algae ecology. *Journal WPCF* **44**, 2294.
SHERMAN, L. A. & HASELKORN, R. (1970). LPP-1 infection of the blue-green alga *Plectonema boryanum.* 1. Electron microscopy. *J. Virol.* **6**, 820.
SHERMAN, L. A. & HASELKORN, R. (1971). Growth of the blue-green algae virus LPP-1 under conditions which impair photosynthesis. *Virology* **45**, 739.
SHILO, M. (1970). Lysis of blue-green algae by Myxobacter. *J. Bact.* **104**, 453.
SHILO, M. (1971). Biological agents which cause lysis of blue-green algae. *Mitt. Internat. Verein. Limnol.* **19**, 206.
SINGH, P. K. (1973). Occurrence and distribution of cyanophages in ponds, sewage and rice fields. *Arch. Mikrobiol.* **89**, 169.
SINGH, R. N. & SINGH, P. K. (1967). Isolation of cyanophages from India. *Nature, Lond.* **216**, 1020.
SINGH, R. N., SINGH, P. K. & VARANASI, P. K. (1969). Lysogeny and induction of lysis of blue-green algae and their viruses. *Proc. 56th Int. Sci. Congr.*
SIRENKO, L. A. (1973). In PADAN, E. & SHILO, M. (1973). Cyanophages – viruses attacking blue-green algae. *Bact. Rev.* **37**, 343.
SKUJA, H. (1948). Taxonomie des Phytoplanktons einiger Seen in Uppland, Schweden. *Symb. bot. upsal.* **9**, 1.
SKUJA, H. (1956). Taxonomische und biologische Studien über das Phytoplankton Schwedischer Binnengewässer. *Nova Acta R. Soc. Scient. upsal.* (Ser. 4) **16**, 1.
SMITH, K. M., BROWN, R. M., GOLDSTEIN, D. A. & WALNE, P. L. (1966a). Culture methods for the blue-green alga *Plectonema boryanum* and its virus, with an electron microscope study of virus infected cells. *Virology* **28**, 580.
SMITH, K. M., BROWN, R. M., WALNE, P. L. & GOLDSTEIN, D. A. (1966b). Electron microscopy of the infection process of the blue-green algal virus. *Virology* **30**, 182.
SMITH, K. M., BROWN, R. M. & WALNE, P. L. (1967). Ultrastructural and time-lapse studies on the replication cycle of the blue-green algal virus LPP-1. *Virology* **31**, 329.
SPARROW, F. K., PATERSON, R. A. & JOHNS, R. M. (1965). Additions to the phycomycete flora of the Douglas Lake Region. V. New or interesting fungi. *Pap. Mich. Acad. Sci.* **50**, 115.
STEWART, J. R. & BROWN, R. M. (1969). *Cytophaga* that kills or lyses algae. *Science N.Y.* **164**, 1523.
STEWART, J. R. & BROWN, R. M. (1970). Killing of green and blue-green algae by a non-fruiting myxobacterium, *Cytophaga* N-5. *Bact. Proc.* p. 18.
STEWART, J. R. & BROWN, R. M. (1971). Algicidal non-fruiting myxobacteria with high G+C ratios. *Arch. Mikrobiol.* **80**, 176.
STEWART, W. D. P. & DAFT, M. J. (1976). Microbial pathogens of Cyanophycean blooms.

In *Advances in Aquatic Microbiology* Eds M. R. Droop & H. Jannasch. London & New York: Academic Press.

STEWART, W. D. P., FITZGERALD, G. P. & BURRIS, R. H. (1967). *In situ* studies on N_2-fixation using the acetylene reduction technique. *Proc. Nat. Acad. Sci. U.S.A.* **58,** 2071.

SZKLARCZYK, C. (1956). Phytoplankton of the dam reservoir at Kzlowa Gora in the years 1951-53. *Acta Soc. Bot. Pol.* **25,** 538.

TOTH, R. & WILCE, R. T. (1972). Virus-like particles in the marine alga *Chorda tomentosa* Lyngbye (Phaeophyceae). *J. Phycol.* **8,** 126.

UEDA, K. (1965). Virus-like structures in the cells of the blue-green alga *Oscillatoria princeps. Expl. cell Res.* **40,** 671.

WU, B., HAMDY, M. K. & HOWE, H, B. (1968). Antimicrobial activity of a myxobacterium against blue-green algae. *Bact. Proc.* GP 14.

ZAVARZINA, N. B. (1962). A lytic agent in cultures of *Chlorella pyrenoidosa,* Prings. *Dokl. Akad. Nauk. SSSR* **137,** 291.

ZAVARZINA, N. B. & PROTSENKO, A. E. (1961). The lysis of *Chlorella pyrenoidosa* Prings. cultures. *Dokl. Akad. Nauk. SSSR* **122,** 840.

ZEHNDER, A. & GORHAM, P. R. (1960). Factors influencing the growth of *Microcystis aeruginosa* Kütz. emend. Elenkin. *Can. J. Microbiol.* **6,** 645.

Microbial Degradation of
Oil and Petrochemicals in the Sea

P. McKenzie* AND D. E. Hughes

*Department of Microbiology, University College, Newport Road,
Cardiff CF2 1TA, Wales*

CONTENTS

1. Introduction

IN ORDER to satisfy world energy demands increasingly vast amounts of crude petroleum and refined products are transported across the oceans. Inevitably, either by accident or during normal tanker operations a certain amount of oil is released into the sea. The quantity of oil entering the sea each year from tankers and from other major sources (for example, oil seepages and industrial wastes) has been estimated at 1.9 million tons (Jeffery, 1972). This figure does not include the massive contributions made by aerial fall-out of unburned hydrocarbons and combustion products, and also hydrocarbons synthesized by the marine biota.

It is apparent that man's activities have in recent years resulted in a great increase in the total input of oil to the marine environment; but what is more important, this tends to be concentrated in certain localities. Nevertheless, the absence of gross pollution on the sea surface and shores indicates that considerable destruction of oil takes place. Mechanisms considered to be responsible include solution (Boylan & Tripp, 1971) and evaporation (Smith &

* Present address: Corrosion and Protection Centre, The University of Manchester Institute of Science and Technology, P.O. Box 88, Manchester M60 1QD, England.

MacIntyre, 1971) of light components, ingestion of oil by marine animals (Parker, Freegarde & Hatchard, 1970; Conover, 1971), and photochemical and microbial oxidation. Only the last 2 processes have any significant capacity for oil destruction, although the other mechanisms may result in an increase in the susceptibility of the oil to degradative attack.

Photochemical oxidation is appreciable only in thin oil films, whereas microbial attack is·most significant when oil is dispersed into the bulk of the sea. Thus it is difficult to estimate the relative effects of the 2 processes under natural conditions. Nevertheless, Freegarde, Hatchard & Parker (1971) have shown that for a film of oil of 2.5 μm thickness, exposed to sunlight for 8 h/day, the rate of photodecomposition can be as high as 0.2 tonne/km^2/day (0.2 g/m^2/day). Estimates of the rate of microbial oxidation vary widely depending on experimental conditions (Floodgate, 1972). For example, using a clear refined mineral oil under nutrient enriched batch culture conditions ZoBell (1964) calculated (on the basis of O_2 uptake, CO_2 production, and microbial growth) that the rate of breakdown was equivalent to about 350 g/m^3/year at 25°. Under similar culture conditions, but with various American crudes dispersed on ignited asbestos, a loss of c. 45% (estimated by extraction and dry weight determination) of an original 1 g of oil was demonstrated from 100 ml of medium after 30 days at 25°. This would be equivalent to a rate of oil oxidation of c. 45 kg/m^3/year.

This paper describes studies concerning the microbial dissimilation of crude oil and hydrocarbons under natural and simulated marine conditions, and also the involvement of various environmental factors in the control of the degradative process.

2. Initial Changes in Pollutant Oil

Oil spilled into the sea rapidly accumulates as a surface slick. Microbial infection occurs immediately, but at this stage any changes which may be caused by the organisms are far outweighed by oil losses due to photo-oxidation and to evaporation and solution of light fractions.

Modification of the physical form of the oil precedes and determines further chemical changes; under the influence of wave action and natural or synthetic surfactants oil emulsions are formed. These may take the form of either a true oil-in-water emulsion or alternatively, particularly after a heavy oil spill, a water-in-oil emulsion (Parker, Freegarde & Hatchard, 1970). Microbial attack on the latter (often referred to as 'chocolate mousse') is severely hindered by the low surface area and generally impermeable nature of the material. In such cases biodegradation can proceed only in the outer layer. Horn, Teal & Backus (1970) and Morris (1971) have reported the common occurrence of lumps of floating oil in the North Atlantic Ocean and the Mediterranean Sea. Such 'tar-balls' often consist of a degraded surface layer surrounding an undegraded core, and most probably represent the final product of microbial attack on 'chocolate mousse'.

In contrast, the production of oil-in-water emulsions ensures that the oil becomes dispersed into the bulk of the sea to provide a vast surface area for microbial attack. To this end, artificial dispersing agents are frequently employed to combat pollutant oil slicks. Such agents may be either 'sinkers' (fine, particulate solids, often of an oleophilic and hydrophobic nature) or more commonly, 'detergents' (synthetic surface active agents). Early detergents, such as those used on the *Torrey Canyon* oil spill, were often highly toxic to the marine biota (Perkins, 1968; Smith, 1968; Shelton, 1971) including the micro-organisms responsible for oil degradation (Beastall, pers. comm.). However, new biodegradable detergents of negligible toxicity are now available. Consequently detergent spraying has become the accepted method of oil slick treatment although criticism of such policy may be justified on two main points. Firstly, detergent treatment increases the concentration of utilizable carbon in an ecosystem already over-stressed by the input of excessive amounts of oil. Secondly, the possibility (which has not yet been fully investigated) that detergents may affect the rate of microbial degradation of oil by direct competition as a source of nutrient carbon.

The use of inorganic sinking agents such as siliconized fuel ash, gypsum residue, or carbonized sand (Brown, 1971) would seem to offer distinct ecological advantages over detergents, but practical difficulties of storage, distribution, and application are considered to limit their use.

3. Microbial Infection of Pollutant Oil

As soon as oil enters the sea it becomes contaminated with a variety of micro-organisms, a number of which can grow on the oil and its degradation products. Most workers consider that bacteria are by far the most important micro-organisms involved in oil biodegradation in the sea. Indeed, of almost 60 pure strains of hydrocarbon-oxidizing micro-organisms isolated in this laboratory from marine and estuarine sources, only one is a yeast, the others being almost all Gram negative bacteria belonging to the genera *Pseudomonas, Achromobacter/Alcaligenes,* and *Flavobacterium.* Other oil-oxidizing bacteria isolated from coastal situations include strains of *Acinetobacter, Mycobacterium, Brevibacterium, Corynebacterium,* and *Arthrobacter* (Soli & Bens, 1972; Atlas & Bartha, 1972a).

Hydrocarbon-oxidizing bacteria are found only in low numbers in the open ocean and in unpolluted coastal waters. For example, 90% of water samples taken near the Isle of Skye (Table 1) were found to contain < 4 hydrocarbon-oxidizers/100 ml (the limit of sensitivity of the counting technique). Numbers of heterotrophs were also typically low. Atlas & Bartha (1973) reported similar levels (2 hydrocarbon-oxidizers/100 ml) in some water samples from Raritan Bay, New Jersey; it is probable that counts of this order of magnitude represent the natural 'background' level of hydrocarbon-oxidizing micro-organisms in waters devoid of pollutant hydrocarbons.

Table 1

Abundance of hydrocarbon-oxidizing and heterotrophic bacteria in seawater samples taken south of the Isle of Skye*

Sampling station	Sampling depth (m)	Bacterial count	
		Hydrocarbon-oxidizers† (per 100 ml)	Heterotrophs‡ (per ml)
1	1	< 4	12
	50	< 4	358
2	1	< 4	97
	50	10	249
3	1	< 4	275
	50	< 4	25
	150	< 4	14
4	1	< 4	64
	50	50	58
5	1	< 4	12
	50	< 4	136
6	1	< 4	41
	50	< 4	12
7	1	< 4	7
	50	< 4	11
	140	< 4	5
8	1	< 4	11
	50	< 4	31
9	1	< 4	13
	50	< 4	18

* Samples taken 17/7/72.
† Estimated by Most Probable Number method (American Public Health Association, 1955) in seawater medium with n-hexadecane, incubated at 10° for 28 days.
‡ Samples membrane filtered and filters incubated on Marine Agar at 10° for 14 days.

In contrast, hydrocarbon-oxidizers are much more common in the seawater and sediments of oil polluted areas. In such situations their numbers may be as high as 10^8–10^9/ml of water or mud (ZoBell & Prokop, 1966; ZoBell, 1969). During a number of cruises in the Bristol Channel (which is subject to heavy tanker traffic and, like many British estuaries, receives considerable volumes of domestic and industrial effluent) we have found that water samples from certain regions of the estuary often contain > 3000 hydrocarbon-oxidizers/100 ml (Fig. 1).

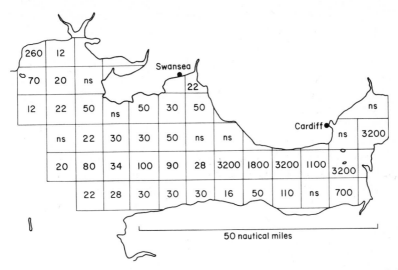

Fig. 1. Abundance and distribution of hydrocarbon-oxidizing micro-organisms in Bristol Channel surface water. Counts are per 100 ml of water. Samples were taken January, 1974. ns, no sample taken.

4. Further Chemical Changes

Microbially infected oils may undergo marked modifications in their chemical composition. The rate at which such changes occur in the sea is influenced by certain environmental factors (see Section 6) and also by the initial composition of the oil.

Crude oil contains 90–99% of hydrocarbons, the remainder comprising compounds containing sulphur, oxygen, and nitrogen, and also trace amounts of metals. The relative concentrations of the hydrocarbon and non-hydrocarbon compounds vary greatly (Brunnock, Duckworth & Stephens, 1968; Dean, 1968) resulting in oils of widely differing properties. The major hydrocarbon components of crude oils are paraffins or alkanes (both straight and short branched chain compounds are found), followed by cycloalkanes (naphthenes) and smaller amounts of aromatic substances containing one or more benzene rings.

Micro-organisms can attack most hydrocarbons found in crude oils and elsewhere (ZoBell, 1964; McKenna & Kallio, 1965). The ability of pure cultures to degrade individual hydrocarbons has been demonstrated with organisms isolated from a wide variety of habitats (Beerstecher, 1954). It has been found that micro-organisms which metabolize n-alkanes are much more frequently encountered than those which degrade naphthenes or aromatic compounds. This has been shown to be the case with organisms of marine and estuarine origin isolated in this laboratory (Byrom, Beastall & Scotland, 1970; Byrom & Beastall, 1971).

The specificity of micro-organisms for hydrocarbon compounds indicates that a mixed population is necessary to accomplish significant destruction of oil. In experiments with mixed cultures and crude oil the *n*-alkane fraction is most readily attacked (Kator, Oppenheimer & Miget, 1971; Atlas & Bartha, 1972*a*; Jobson, Cook & Westlake, 1972) while, as before, naphthenic compounds are more refractory and aromatics the least susceptible (ZoBell, 1946). The progressive degradation of *n*-alkanes during a typical experiment is illustrated in Fig. 2. Low molecular weight compounds are the first to be lost followed by the alkanes of longer chain length until only peaks 17 and 18 remain. At this stage the 2 peaks represent terpenoid compounds (probably pristane and phytane) originally masked by the peaks of *n*-heptadecane and *n*-octadecane; although more resistant than the alkanes, these too are eventually degraded.

Gas chromatography provides a rapid and efficient technique for monitoring

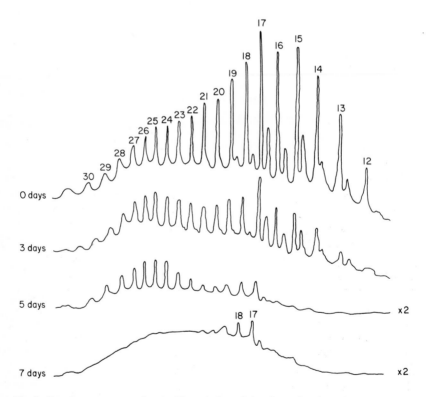

Fig. 2. Gas chromatograms showing degradation of the alkane fraction of Brega crude oil by hydrocarbon-oxidizing micro-organisms at 25°. Seawater medium (100 ml) was incubated with 1 g of oil and 0.2 g of estuarine mud and incubated on an orbital shaker. Numbers above the peaks refer to the *n*-alkane chain length.

changes in the alkane fraction during the biodegeneration of crude oil. However, the difficulties inherent in the analysis of other oil fractions prompted us to utilize model artificial oils to investigate the pattern of degradation of typical crude oil components (Table 2). The use of such hydrocarbon mixtures may also reveal co-oxidation reactions where compounds resistant to attack are degraded in the presence of another hydrocarbon; the reviews of Raymond, Jamison & Hudson (1971) and Horvath (1972) refer to many such cases.

Table 2
Degradation of a hydrocarbon mixture by oil-oxidizing micro-organisms*

Hydrocarbon type	Compound	Percent breakdown†
n-alkane	Nonane	100
	Decane	100
	Dodecane	95.1
	Hexadecane	59.2
	Pentadecane	48.8
	Octadecane	46.3
	Eicosane	44.2
	Tetracosane	36.2
Aromatic/cycloalkane	Tetralin	31.0
Iso-alkane	Pristane	25.4
Aromatic	2-Methyl naphthalene	17.1
Cycloalkane	Dicyclohexyl	15.4
Cycloalkane	Decalin	13.6
Aromatic	Thianaphthalene	13.6
Aromatic	Acenaphthalene	6.1

* Stoppered bottles (500 ml) contained 100 ml of seawater medium and 0.1 ml hydrocarbon mixture (containing equal weights of each compound) inoculated with 0.1 ml Kuwait crude oil enrichment culture. Incubation was in an orbital shaker at 22° for 21 days.
† Estimated by gas chromatography.

Results indicate that the biodegradation of the aromatic fraction of oils may proceed only after the loss of the less resistant components. This fraction contains a small proportion of polycyclic hydrocarbons, a number of which are potent carcinogenic agents, for example, 3,4-benzpyrene which is normally present at 400–1600 μg/kg oil (ZoBell, 1971). In recent years the discovery of polycyclic hydrocarbons in organisms of the marine food chain has aroused considerable attention and concern (Andelman & Suess, 1970; Suess, 1970; ZoBell, 1971). The origin of such compounds is by no means well-defined, although an obvious potential source is pollutant oil. Our studies have shown that normal oil-degrading microflora are capable of decreasing the concentration of low levels of polycyclic hydrocarbons in culture fluids (Table 3). Further investigations with artificial oil mixtures may go some way to predicting the fate of polycyclic hydrocarbons in degrading oils at sea.

Table 3

Degradation of polycyclic aromatic hydrocarbons by oil-oxidizing micro-organisms*

Compound	Concentration (μg/l)†		Percent decrease
	initially	after 12 days	
3,4-Benzpyrene	190	90	53
Pyrene	365	55	85
Fluorene	350	270	23

* Hydrocarbon mixture (in 1 ml cyclohexane) was added to 100 ml seawater medium inoculated with washed organisms from 10 ml Kuwait oil enrichment culture. Cultures incubated stirred at 10°.
† Estimated by UV spectrophotometry.

5. Biochemistry of Hydrocarbons

The reactions involved in hydrocarbon catabolism are well documented, and have been summarized by Davis & Hughes (1968). Of particular note is the involvement of molecular oxygen in the primary reactions; the degradation of both alkanes and aromatic compounds requires the addition of molecular oxygen under the influence of oxygenase enzymes. In the case of alkanes, subsequent reactions give rise to fatty acids which are fed into the β-oxidation spiral. Acetyl coenzyme A produced during the sequential degradation of the fatty acids is metabolized via the tricarboxylic acid cycle. The addition of oxygen to aromatic compounds results in ring fission followed by reactions leading, as before, to acetyl coenzyme A and also to intermediates of the tricarboxylic acid cycle.

The completed reactions would, in theory, give rise to the end products CO_2, water, biomass, and energy. In practice, many incompletely oxidized products accumulate including organic acids, esters, alcohols, ketones, aldehydes, and others (McKenna & Kallio, 1965; Davis, 1967).

ZoBell (1964) estimated that 25–35% of the carbon content of alkanes was converted into biomass. Such microbial substance may contribute significantly to the nutrition of marine animals.

6. Factors Governing the Rate of Oil Degradation

(a) *Temperature*

The temperature of the environment is an important factor governing microbial activity. At temperatures below their optimum, the generation time of most bacteria is decreased 2–3 fold by a 10° rise.

The temperature of the sea ranges from about $-2°$ to about $30°$ but, by volume, c. 90% is cooler than $5°$. Most laboratory studies of oil degradation have involved incubation temperatures of $20-37°$ resulting in the selection and enrichment of organisms growing in these temperature ranges. However, the oil pollution problem is often more severe at lower temperatures such as prevail in the North Sea and the North Alaska slope. Under such conditions mesophilic micro-organisms would have only a small effect on oil dissimilation.

Psychrophilic micro-organisms, which grow at low temperatures, are frequently found in the deep sea (ZoBell, 1968). ZoBell & Agosti (1972) have found bacteria which can degrade oil in ice-water between $-1.5°$ to near $0°$. Such organisms are very sensitive to increased temperatures; one bacterium isolated at temperatures below $8°$ failed to grow at $18°$ and was killed within 10 min at $25°$. Similarly, we have found that many hydrocarbon-oxidizing bacteria isolated at $10°$ from Loch Etive, Argyllshire grow well at $15°$ but not at all at $25°$. These observations emphasize the care needed in the isolation of such organisms.

Besides affecting the rate of oil degradation, changes in temperature may also indirectly alter the pattern of hydrocarbon utilization; under cold conditions the precipitation from crude oil of certain alkanes as waxes would greatly diminish their availability to oil-degrading organisms.

(b) Oxygen

The biodegradation of alkanes and aromatic compounds is absolutely dependent on a supply of molecular oxygen; the complete oxidation of 1 mg of hydrocarbon requires 3–4 mg of oxygen. The normal concentrations of dissolved oxygen found in surface waters (6–12 mg/l) are probably quite adequate to support microbial growth on thin oil slicks. However, in regions of intense microbial activity such as bottom sediments, or under conditions of heavy oil pollution, oxygen may be consumed faster than it can be replenished by diffusion, water turbulence, and photosynthesis. For example, after a severe fuel oil spill in Buzzard's Bay, Massachusetts, oil which entered the sediment remained virtually unchanged for several months; only near the sediment surface had any significant degradation taken place (Blumer, Souza & Sass, 1970; Blumer & Sass, 1972). Similarly, in experiments with 'heavily oiled' sand columns (1.1 kg oil/m^2), the dissolved oxygen content was found to decrease rapidly in the interstitial water below the oil; only c. 10% of the added oil was degraded, and the remainder decayed 'immeasurably slowly' (Johnston, 1970).

A number of claims have been made for the anaerobic dissimilation of alkanes (McKenna & Kallio, 1965; ZoBell & Prokop, 1966; Traxler & Bernard, 1969), but such reactions constitute at best an extremely slow degradative mechanism.

(c) *Nitrogen and phosphorus*

Unpolluted seawater contains only very low levels of nitrogen- and phosphorus-containing compounds. In the presence of excess utilizable carbon (as in the case of an oil spill), it is probable that such low mineral nutrient concentrations are responsible for limitation of microbial growth and oil biodegradation.

In laboratory investigations seawater/oil media are commonly enriched with phosphate and a source of nitrogen (usually ammonium or nitrate) in order to obtain maximum growth of oil-degrading micro-organisms. For example, Atlas & Bartha (1972*a*) demonstrated only 3% breakdown of the alkane fraction of oil added to natural seawater, compared with 70% after phosphate and nitrate enrichment. In similar experiments (Table 4) we have found that nutrient enrichment of surface water samples from the Bristol Channel greatly increases

Table 4

The effect of nitrogen and phosphorus addition on the biodegradation of oil in surface water from the Bristol Channel*

Sampling station	Percent oil loss‡	
	without added N and P	with added N and P†
10	0	55
23	0	52
25	8	9
27	3	27
33	9	50
35	0	53
37	0	5
49	1	17

* 1 g Kuwait crude oil was added to 100 ml of natural seawater, incubated in an orbital shaker at 25° for 21 days.
† 0.1 g NH_4Cl and 0.05 g K_2HPO_4.
‡ Estimated by dry weight determination of residual oil following extraction in diethyl ether.

the extent of oil degradation, although the spread of results indicates that other factors (possibly interference by alternative carbon sources and heavy metals) may be important in controlling the oil degradation process in this estuary. Thus it is apparent that even in this watercourse, which is subject to gross organic pollution, nutrient levels are inadequate for sustained biodegradation under conditions of chronic oil pollution.

Atlas & Bartha (1972*b*) have considered the possibility of addition of mineral nutrients to oil slicks at sea. The rapid dilution of nitrogen and phosphorus salts in the bulk of the sea, and the possibility of the encouragement of algal blooms would seem to prohibit such action.

Bacteria have been isolated which fix atmospheric nitrogen and oxidize hydrocarbons (Davis, Coty & Stanley, 1964; Coty, 1967). Such organisms may have great potential value for 'seeding' oil slicks, although this possibility has not been investigated.

(d) *Alternative carbon sources*

The concentration of dissolved organic matter in unpolluted areas of the sea rarely exceeds 1–2 mg/l (Wagner, 1969). However, in certain coastal regions, domestic and industrial effluents introduce large amounts of a wide variety of organic substances into the sea. Such materials may interfere with the biodegradation of hydrocarbons.

ZoBell (1946) noted that low concentrations of organic matter ($<$ 1 mg/l) generally promote microbial action on hydrocarbons, probably by effecting an initial rapid growth of the organisms. We have found that higher concentrations of alternative carbon sources such as glucose, or fatty acids result in a decrease in the rate of oil oxidation (Hughes & McKenzie, 1975). Thus, under certain conditions, fatty acid by-products of the oil oxidation system may play a role in the overall regulation of the process.

The possibility that oil spill detergents may compete with oil as a source of available carbon is at present under investigation.

7. Experiments under Natural and Simulated Field Conditions

The practical difficulties inherent in confining oil at sea have resulted in a general neglect of field experiments; the majority of available data concerns observations made following accidental spillages.

Our group has been fortunate in being able to carry out a number of experiments under natural conditions; a lagoon at Aberthaw and a dock at Swansea have been used to study the fate of sunken oil (Hughes & McKenzie, 1975), and also the biodegradation of floating and detergent-treated oil has been investigated in Langstone Harbour, Portsmouth (Beastall & Hughes, 1976).

(a) *Sunken oil in Loch Etive*

Further experiments with sunken oil have been conducted in Loch Etive (Argyllshire), a marine loch which has not been subject to any significant oil pollution. Two experimental stations were used. Both had a depth of *c.* 30 m and a mud bottom sediment, but contrasting levels of oxygenation were apparent near the sea bed; the seawater at Station I, near the mouth of the loch, was almost fully oxygen-saturated, whereas that at Station II, in the upper reaches of the loch, usually exhibited 10–20% saturation.

Sunken oil test samples were prepared by mixing siliconized fuel ash or 'Nautex-H' (stearated chalk whiting; Wolon Co. Ltd.) sinker with fresh crude oil (Brega or Kuwait) in the proportions 2 : 1 and 3 : 1 (w/w) respectively to the oil. Duplicate mixes were supplemented with agricultural fertilizers (10% Basic Slag and 5% Nitro Chalk relative to the oil) to provide a readily available source of nitrogen and phosphorus. Divers deposited the sunken oil samples (usually 4 or 8 kg) on the sea bed at the two stations. The samples formed a layer of c. 3–4 cm thickness and were enclosed within mesh-covered retaining frames for ease of location.

Core samples of the oil and underlying sediment were retrieved at intervals over the following year. Microbial numbers were estimated by plate counts made on samples from the oil surface and on mud from the oil-sediment interface. To produce the initial decimal dilution 1.0 g of sunken oil was emulsified with 4.5 ml of Corexit and 4.5 ml of artificial seawater (Rila Products, New Jersey). Heterotrophs were counted on Marine agar (Difco) and hydrocarbon-oxidizers were estimated by the Most Probable Number method (*American Public Health Association*, 1955) using artificial seawater supplemented with $(NH_4)_2 HPO_4$ (0.1% w/v) and *n*-hexadecane (0.1% v/v).

Figure 3 illustrates the typical microbial growth pattern on sunken oil in Loch Etive. At both stations rapid colonization of the oil occurred during the first month of the experiment. Maximum bacterial growth (generally c. 10^6/g oil

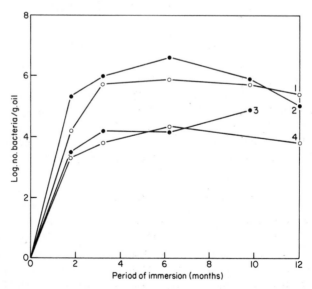

Fig. 3. Bacterial growth on sunken oil in Loch Etive. Curves 1 and 2, heterotrophic bacteria. Curves 3 and 4, hydrocarbon-oxidizing bacteria, ○, oil without added nutrients; ●, oil with nutrients.

for heterotrophs, and *c*. 10^4/g for hydrocarbon-oxidizers) was recorded within 8–24 weeks after submersion and was usually followed by a gradual decline in numbers. All 4 combinations of oils and sinkers supported similar numbers of bacteria; the inclusion of fertilizers had little effect but, in some cases led to marginally higher bacterial counts.

Bacterial growth in the mud at the oil/sediment interface is shown in Fig. 4. As before the pattern and levels of growth were similar at the 2 stations. During the year of submersion the numbers of both heterotrophs and hydrocarbon-oxidizing bacteria were generally greater in mud underlying the oil samples than in non-oiled mud; this was presumably a reaction to sparingly soluble hydrocarbons leaching from the oil. The sediment bacteria exhibited no consistent reaction to nutrients leaching from the fertilizer-treated oils.

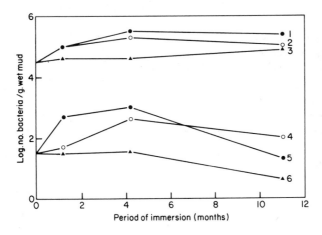

Fig. 4. Bacterial growth in mud below sunken oil in Loch Etive. Curves 1, 2 and 3, heterotrophic bacteria. Curves 4, 5 and 6, hydrocarbon-oxidizing bacteria. ○, mud below oil without added nutrients; ●, mud below oil with added nutrients; ▲, control mud samples.

Sunken oil recovered after almost one year of submersion was found to be unchanged in appearance, and still smelled 'fresh' when squashed. Gas chromatography showed that little degradation had occurred; only light alkanes up to undecane had been lost from the oil surface. Nevertheless, it was found that the sunken oil, when inoculated into a floating oil/seawater medium, produced extensive degradation of the fresh oil within 3 days at 15°. It was concluded that biodegradation of the oil in Loch Etive had been severely limited by the low surface area of the sunken oil mass. Laboratory studies have shown that extensive bacterial growth is possible only in the surface layer of sunken oil; penetration of the oil mass is accomplished only after degradation of this outer layer.

For the Loch Etive experiments large, compact sunken oil mixes were used in order to facilitate sampling. In practice, the sinking of an oil slick should result in smaller lumps and particles of oil-coated sinker. This knowledge, together with observations that the oil mass gradually sank into, or became silted over by the sediment of Loch Etive, prompted an investigation of the fate of oil in bottom sediment.

(b) *Sunken oil in marine sediment*

Under natural conditions it is probable that the ultimate degradative attack of pollutant oil occurs within the sediment. Nevertheless, few attempts have been made to estimate the rate of oil biodegradation in marine silts. Our experiments have involved the use of sediment columns in order to simulate natural conditions.

To prepare the columns, surface silt (from the bed of Loch Etive, near Station II) was mixed with sunken oil (siliconized fuel ash plus Brega oil, pre-weathered at laboratory temperature) to give *c.* 1 g oil/100 g of wet sediment. The mixture was packed into Perspex tubes (26 x 5.5 cm) which were closed at one end, and incubated upright under flowing seawater. The water temperature varied between 9° and 13°.

Redox potential (E_h) measurements were made directly down the length of

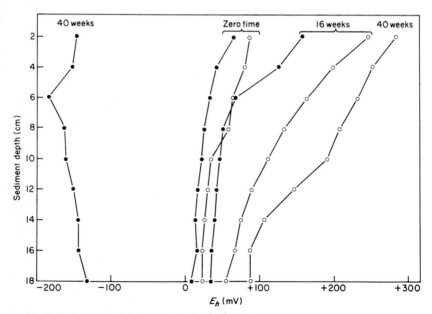

Fig. 5. Redox potential changes in oiled and non-oiled model sediment columns. •, oiled; ○, non-oiled.

the columns 16 and 40 weeks after immersion (Fig. 5). In control (non-oiled) columns the E_h values became progressively more positive over the period of the experiment. In contrast, conditions in the oiled columns after 16 weeks were significantly more reducing than in the controls. After 40 weeks, E_h values in the oiled cores were extremely low, and sulphide blackening of the mud was apparent due to the growth of sulphate-reducing bacteria in the oxygen-depleted conditions.

Gas chromatography of residual oil extracted after 40 weeks revealed extensive degradation in the upper 12 cm of the columns (Fig. 6). Below this level the oil remained virtually unchanged.

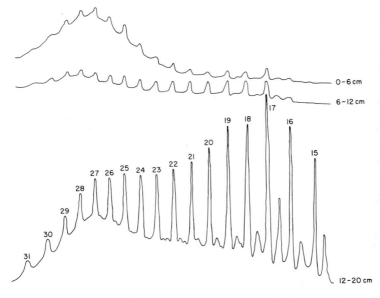

Fig. 6. Gas chromatograms of residual oil at various depths in a model sediment system. The oil was extracted in diethyl ether after 10 months of incubation.

If it is assumed that the rate of biodegradation of the alkanes far exceeds that of the other oil fractions, then measurement of the residual alkanes facilitates an approximation of the overall rate of oil breakdown. By measurement of the chromatogram peak areas and comparison with the original (time zero) values, it is apparent that a total of 62% of the alkanes were degraded within the oiled column. The column contained initially 5.50 g of oil, of which the alkanes comprised 8.7% (0.48 g). Thus 0.30 g of alkanes was lost during the experiment. Taking into account the column dimensions and the period of incubation, the rate of oil breakdown is estimated at $0.30 \, \text{g}/0.0025 \, \text{m}^2/316$ days, or

0.38 g/m^2/day. This rate may be compared with estimates made by Johnston (1970) using oiled sand columns (0.04–0.09 g/m^2/day, depending on dosage).

Although such results seem encouraging, it should be emphasized that these rates of oil biodegradation are applicable only for a relatively short period of time; that is, until oxygen depletion of the sediment precludes further microbial attack on the oil. Under such conditions, it is likely that any residual oil will remain unchanged for long periods of time.

8. Acknowledgements

The support of the Natural Environment Research Council and the Department of Trade and Industry is gratefully acknowledged.

9. References

AMERICAN PUBLIC HEALTH ASSOCIATION (1955). *Standard Methods for the Examination of Water, Sewage, and Industrial Wastes.* New York: A.P.H.A.

ANDELMAN, J. B. & SUESS, M. J. (1970). Polynuclear aromatic hydrocarbons in the water environment. *Bull. Wld. Hlth. Org.* **43**, 479.

ATLAS, R. M. & BARTHA, R. (1972*a*). Degradation and mineralization of petroleum by two bacteria isolated from coastal waters. *Biotechnol. & Bioeng.* **14**, 297.

ATLAS, R. M. & BARTHA, R. (1972*b*). Degradation and mineralization of petroleum in sea water: Limitation by nitrogen and phosphorus. *Biotechnol. & Bioeng.* **14**, 309.

ATLAS, R. M. & BARTHA, R. (1973). Abundance, distribution and oil biodegradation potential of microorganisms in Raritan Bay. *Environ. Pollut.* **4**, 291.

BEASTALL, S. & HUGHES, D. E. (1976). Microbial breakdown of crude oil. *J. appl. Chem. Biotechnol.* (In press.)

BEERSTECHER, E. (1964). *Petroleum Microbiology.* New York: Elsevier Press.

BLUMER, M. & SASS, J. (1972). Indigenous and petroleum derived hydrocarbons in a polluted sediment. *Mar. Pollut. Bull.* **3**, 92.

BLUMER, M., SOUZA, G. & SASS, J. (1970). Hydrocarbon pollution of edible shellfish by an oil spill. *Mar. Biol.* **5**, 195.

BOYLAN, D. B. & TRIPP, B. W. (1971). Determination of hydrocarbons in seawater extracts of crude oil and crude oil fractions. *Nature, Lond.* **230**, 44.

BROWN, R. B. (1971). Treatment of oil spills by sinking: A review of requirements and progress. *J. Inst. Petrol.* **57**, 8.

BRUNNOCK, J. V., DUCKWORTH, D. F. & STEPHENS, G. C. (1968). Analysis of beach pollutants. In *Scientific Aspects of Pollution of the Sea by Oil.* London: Institute of Petroleum.

BYROM, J. A. & BEASTALL, S. (1971). Microbial degradation of crude oil with particular emphasis on pollution. In *Microbiology, 1971.* London: Institute of Petroleum.

BYROM, J. A., BEASTALL, S. & SCOTLAND, S. (1970). Bacterial degradation of crude oil. *Mar. Pollut. Bull.* **1**, NS(2), 25.

CONOVER, R. J. (1971). Some relations between zooplankton and Bunker C oil in Chedabucto Bay following the wreck of the tanker 'Arrow'. *J. Fish. Res. Bd. Can.* **28**, 1327.

COTY, V. F. (1967). Atmospheric nitrogen fixation by hydrocarbon oxidizing bacteria. *Biotechnol. & Bioeng.* **9**, 25.

DAVIES, J. A. & HUGHES, D. E. (1968). The biochemistry and microbiology of crude oil degradation. *Fld. Stud.* **2** (Suppl.), 139.

DAVIS, J. B. (1967). *Petroleum Microbiology.* London: Elsevier Press.

DAVIS, J. B., COTY, V. F. & STANLEY, J. P. (1964). Atmospheric nitrogen fixation by methane oxidizing bacteria. *J. Bact.* **88**, 468.

DEAN, R. A. (1968). Chemistry of crude oils in relation to their spillage on the sea. *Fld. Stud.* **2** (Suppl.), 1.

FLOODGATE, G. D. (1972). Microbial degradation of oil. *Mar. Pollut. Bull.* **3**, 41.

FREEGARDE, M., HATCHARD, C. G. & PARKER, C. A. (1971). Oil spilt at sea: its identification, determination and ultimate fate. *Lab. Pract.* **20**, 35.

HORN, M. H., TEAL, J. M. & BACKUS, R. H. (1970). Petroleum lumps on the surface of the sea. *Science, N.Y.* **168**, 245.

HORVATH, R. S. (1972). Microbial co-metabolism and the degradation of organic compounds in Nature. *Bact. Rev.* **36**, 146.

HUGHES, D. E. & McKENZIE, P. (1975). The microbial degradation of oil in the sea. *Proc. R. Soc. Lond. B.* **189**, 375.

JEFFERY, P. G. (1972). *Oil in the Marine Environment.* Warren Spring Lab. Rept. LR156(PC).

JOBSON, A., COOK, F. D. & WESTLAKE, D. W. S. (1972). Microbial utilization of crude oil. *Appl. Microbiol.* **23**, 1082.

JOHNSTON, R. (1970). The decomposition of crude oil residues in sand columns. *J. mar. biol. Ass. U.K.* **50**, 925.

KATOR, H., OPPENHEIMER, C. H. & MIGET, R. J. (1971). Microbial degradation of a Louisiana crude oil in closed flasks and under simulated field conditions. *Proc. Joint Conf. Prevention and Control of Oil Spills* (A.P.I., E.P.A. & U.S.C.G., June, 1971, Washington, D.C.). New York: A.P.I.

McKENNA, E. J. & KALLIO, R. E. (1965). The biology of hydrocarbons. *Ann. Rev. Microbiol.* **19**, 193.

MORRIS, B. F. (1971). Petroleum: tar quantities floating in the northwestern Atlantic taken with a new quantitative neuston net. *Science, N.Y.* **173**, 430.

PARKER, C. A., FREEGARDE, M. & HATCHARD, C. G. (1970). The effect of some chemical and biological factors on the degradation of crude oil at sea. In *Water Pollution by Oil.* London: Institute of Petroleum.

PERKINS, E. J. (1968). Toxicity of oil emulsifiers to some inshore fauna. *Fld. Stud.* **2** (Suppl.), 81.

RAYMOND, R. L., JAMISON, V. W. & HUDSON, J. O. (1971). Hydrocarbon cooxidation in microbial systems. *Lipids* **6**, 453.

SHELTON, R. G. J. (1971). Effects of oil and oil dispersants on the marine environment. *Proc. R. Soc. Lond. B.* **177**, 411.

SMITH, C. L. & MacINTYRE, W. G. (1971). Initial aging of fuel oil films of sea water. *Proc. Joint Conf. Prevention and Control of Oil Spills* (A.P.I., E.P.A. & U.S.C.G., June, 1971, Washington, D.C.). New York: A.P.I.

SMITH, S. E. (1968). *Torrey Canyon Pollution and Marine Life.* Cambridge University Press.

SOLI, G. & BENS, E. M. (1972). Bacteria which attack petroleum hydrocarbons in a saline medium. *Biotechnol. & Bioeng.* **14**, 319.

SUESS, M. J. (1970). Occurrence of polycyclic aromatic hydrocarbons in coastal waters and their possible effect on human health. *Arch. Hyg. Bakteriol.* **154**, 1.

TRAXLER, R. W. & BERNARD, J. M. (1969). The utilization of n-alkanes by *Pseudomonas aeruginosa* under conditions of anaerobiosis. 1. Preliminary observation. *Int. Biodetn. Bull.* **5**, 21.

WAGNER, F. S. (1969). Composition of the dissolved organic compounds in seawater: a review. *Contr. Mar. Sci. Univ. Tex.* **14**, 115. (*via Wat. Pollut. Abstr.* 1970, **43**, abstr. no. 1319).

ZoBELL, C. E. (1946). Action of microorganisms on hydrocarbons. *Bact. Rev.* **10**, 1.

ZoBELL, C. E. (1964). The occurrence, effects, and fate of oil polluting the sea. *Adv. Wat. Pollut. Res.* **3**, 85.

ZoBELL, C. E. (1968). Bacterial life in the deep sea. *Bull. Misaki Mar. Biol. Inst., Kyoto Univ.* No. 12, 77.

ZoBELL, C. E. (1969). Microbial modification of crude oil in the sea. *Proc. Joint Conf. Prevention and Control of Oil Spills* (A.P.I. & F.W.P.C.A., December, 1969, New York.). New York: A.P.I.

ZoBELL, C. E. (1971). Sources and biodegradation of carcinogenic hydrocarbons. *Proc. Joint Conf. Prevention and Control of Oil Spills* (A.P.I., E.P.A. & U.S.C.G., June, 1971, Washington, D.C.). New York: A.P.I.

ZoBELL, C. E. & AGOSTI, J. (1972). Bacterial oxidation of mineral oils at sub-zero Celsius. *Bacteriol. Proc.* E 11.

ZoBELL, C. E. & PROKOP, J. F. (1966). Microbial oxidation of mineral oils in Barataria Bay bottom deposits. *Zeit. Allg. Mikrobiol.* **6**, 143.

Trends in Silage Making

P. McDONALD

*Department of Agricultural Biochemistry, Edinburgh School of Agriculture,
University of Edinburgh, West Mains Road, Edinburgh EH9 3JG, Scotland*

CONTENTS

1. Introduction

DURING the past decade, a considerable amount of research has been carried out to improve our understanding of the microbiological and biochemical changes which occur during the conservation of grass as silage. Previous reviews of this subject are those of Watson & Nash (1960), Whittenbury (1968), Ohyama (1971), Zimmer (1971), Woolford (1972), McDonald & Whittenbury (1973) and Wilkins (1974). This present review summarizes the main changes which occur during the natural fermentation of grass by lactic acid bacteria and discusses the effect of these changes upon subsequent utilization of silage by ruminant animals. Methods of manipulating the fermentation in order to obtain a more acceptable product are described.

2. Biochemical Changes during Ensilage

The water soluble carbohydrates of grasses include glucose, fructose, sucrose and fructans, but for microbiological purposes we can consider that glucose and fructose are the main carbohydrate sources. The sugars surviving aerobic metabolism are fermented by many types of micro-organisms although lactic acid bacteria soon become dominant and rapidly inhibit unwanted bacteria. Table 1 lists those lactic acid bacteria most commonly found on fresh herbage and in silage (Gibson *et al.*, 1958; Langston & Bouma, 1960; Gibson *et al.*, 1961).

Lactic acid bacteria are relatively scarce on fresh crops (Stirling, 1953; Kroulik, Burkey & Wiseman, 1955; Gibson *et al.*, 1958), their numbers often

Table 1

*Some species of lactic acid bacteria commonly found
on fresh herbage and in silage*

Homofermentative	Heterofermentative
Lactobacillus plantarum	*Lactobacillus brevis*
Pediococcus acidilactici	*Lactobacillus buchneri*
Streptococcus durans	*Lactobacillus fermentum*
Streptococcus faecalis	*Lactobacillus viridescens*
Streptococcus faecium	*Leuconostoc mesenteroides*
Streptococcus lactis	

being $< 100/g$ fresh grass. The number of these organisms found on fresh herbage is influenced by the quantity of decayed material near the ground, since this is their usual location on the plant (Stirling & Whittenbury, 1963). The forage harvester, which lacerates or chops the grass, undoubtedly plays an important role in the inoculation of herbage with lactic acid bacteria (see Table 2).

Table 2

*Microbial counts (no. of organisms/g fresh material) on fresh
grass and silage samples*

	Untreated grass	
	Total count	Lactic acid bacteria
Uncut grass	5.9×10^6	< 100
Forage harvested grass	2.5×10^8	4.9×10^5
Grass at silo	3.0×10^8	8.3×10^4
Silage (after 189 days)	3.9×10^4	7.0×10^3

The fermentation of sugars is variable depending upon whether the homofermentative or heterofermentative lactic acid bacteria are dominant, the former being more efficient at producing lactate from hexoses (Table 3). In silages made directly from fresh herbage, the residual sugar levels are extremely low, usually 1–2% of the dry matter. Pentoses are often the commonest components of these residual sugars, being derived from hemicellulose hydrolysis (Dewar, McDonald & Whittenbury, 1963). Apart from sugars, other components in the herbage are changed during ensilage. The organic acids, citrate and malate are fermented by a number of pathways resulting in the formation of several products including lactate, acetate, formate, ethanol, 2,3-butanediol and acetoin (McDonald & Whittenbury, 1973). Organic acids play an important role as buffering agents in plants. Within the pH range 4–6 *c.* 68–80% of the buffering capacity of herbage can be attributed to the anions (organic acid salts, orthophosphates, sulphates, nitrates and chlorides), with only *c.* 10–20% resulting from the action of plant proteins (Playne & McDonald, 1966).

The nitrogenous components of herbage consist mainly of protein which is rapidly hydrolysed to amino acids immediately after harvesting (MacPherson & Slater, 1959). Further breakdown of individual amino acids may occur, the extent depending upon the rate of pH fall (MacPherson & Violante, 1966). Arginine and serine are attacked by the lactic acid bacteria, these being deaminated and decarboxylated to ornithine and acetoin respectively (Whittenbury *et al.*, 1967). The situation is entirely different if clostridia become dominant, which sometimes happens if crops are ensiled too wet or an initial low pH has not been achieved (Weiringa, 1958; Gibson, 1965). The classical picture of a clostridial fermented silage is an unstable product of relatively high pH containing significant amounts of butyric acid, branch-chain fatty acids and ammonia (Table 3).

Table 3
Main products of fermentation of some nutrients by silage bacteria

Homofermentative lactic acid bacteria

Glucose	→	2 Lactate
Fructose	→	2 Lactate
Pentose	→	Lactate and Acetate
2 Citrate*	→	Lactate and 3 Acetate + 3 CO_2
Malate*	→	Lactate + CO_2

Heterofermentative lactic acid bacteria

Glucose	→	Lactate + Ethanol + CO_2
3 Fructose	→	Lactate + 2 Mannitol + Acetate + CO_2
2 Fructose + Glucose	→	Lactate + 2 Mannitol + Acetate + CO_2

Clostridia

2 Lactate	→	Butyrate + 2 CO_2 + 2 H_2
Alanine + 2 Glycine	→	3 Acetate + 3 NH_3 + CO_2
3 Alanine	→	2 Propionate + Acetate + 3 NH_3 + CO_2
Valine	→	*Iso*butyrate + NH_3 + CO_2
Leucine	→	*Iso*valerate + NH_3 + CO_2
Histidine	→	Histamine + CO_2
Lysine	→	Cadaverine + CO_2
Arginine	→	Putrescine + 2 CO_2 + 2 NH_3

* Pathways are similar for heterofermentative lactic acid bacteria.

3. Efficiency of Ensilage as a Conservation Technique

If lactic acid fermentation is considered *per se,* it is extremely efficient in terms of dry matter recovery; even a heterolactic fermentation is unlikely to result in losses > 3–6%. Such losses which may be predicted from biochemical calculations (McDonald & Whittenbury, 1973), have also been obtained experimentally by several workers (Brown & Kerr, 1965; Anderson & Jackson, 1969).

Dry matter recovery, however, may be a misleading measurement of conservation efficiency and a consideration of the energy changes might be more appropriate. For example, in considering the heterolactic fermentation of glucose, dry matter recovery is only 76% but energy recovery is c. 98%, because of the formation of the high energy compound ethanol. This shift during fermentation towards high energy, reduced compounds results in a concentration of energy in silage. Increases of up to 10% in gross energy concentrations have been reported in laboratory silages (McDonald, Henderson & Ralton, 1973).

Losses of nutrients during ensilage arise from the production of effluent from the silo (McDonald et al., 1962). The flow of effluent is related to the dry matter content of the crop ensiled and Zimmer (1967) has produced the regression equation:

$$W = 83.26 - 5.418\,D + 0.0883\,D^2$$

where W is the weight of effluent as % of fresh herbage and D is the % dry matter of the ensiled herbage. It is clear from the findings of Zimmer (1967) and others (McDonald & Whittenbury, 1973), that crops ensiled at c. 30% dry matter produce little if any effluent. In the field situation, dry matter and gross energy losses much higher than those quoted above occur; these arise mainly from oxidation.

4. Influence of Oxygen on Ensilage

If anaerobiosis is not achieved in the silo, plant cells continue to respire resulting in the oxidation of soluble sugars with the liberation of heat

$$C_6H_{12}O_6 + 6\,O_2 \rightarrow 6\,CO_2 + 6\,H_2O + 2815\ \text{kJ}$$

Such depletion of sugars clearly results in less being available for subsequent fermentation but in addition, delayed anaerobiosis can also change the relative counts of different micro-organisms (Takahashi, 1970). Lactic acid bacteria, being facultative anaerobes, should be favoured by oxygen, compared with obligate anaerobes such as clostridia. At high levels of oxygen, Takahashi (1970) reported that the effect was obscured by an opposing factor, yeast development, which may inhibit lactic acid production. Weise (1968) has confirmed that yeasts are stimulated by oxygen in the early stages of ensilage. Yeast multiplication makes no contribution to preservation and in that sense is a disadvantage. Ohyama et al. (1970) reported that aerobiosis encouraged the development of Gram negative bacteria. Such micro-organisms of the coliform type produce a variety of fermentation products including lactate, acetate, ethanol, 2,3-butanediol, CO_2 and H_2 and make only a minor contribution to fall in pH. The general effect of depletion of sugars in the presence of high oxygen concentrations is to encourage a clostridial fermentation as shown in Table 4 in which experiment grass was ensiled in laboratory silos with different volumes of oxygen (Ruxton & McDonald, 1974).

5. Effect of Fermentation Changes on the Nutritional Value of Silage

The nutritional value of a food depends primarily upon the extent to which it is digested and secondly on the efficiency with which the digested nutrients are utilized by the animal for maintenance and production. A third factor, of equal importance, is the quantity of food that an animal can eat in a given time, generally expressed in terms of intake of dry matter/day in relation to some function of the animal's liveweight (W) such as $W^{0.75}$.

Our own studies at Edinburgh with a wide range of silages, indicate that the digestibility of well-preserved (lactate dominant) silages is similar to that of the original grass.

We have much less information about the efficiency of utilization of digested nutrients although it is generally accepted that the type of fermentation in the rumen can influence its efficiency, especially for lipogenesis. Ruminal end-products of digestion in which acetate predominates are used much less efficiently than those richer in propionate (Blaxter, 1967). Diets rich in soluble carbohydrates tend to result in the production of relatively high proportions of propionate in the rumen and it might be reasonable to assume that because of low residual sugar contents, normal silages would result in a high acetate ruminal fementation. There is evidence for this assumption from the *in vivo* studies of Anderson & Jackson (1971) who obtained slightly higher ruminal acetate levels on silage diets compared with fresh grass and more recently results at Edinburgh (Donaldson & Edwards, unpublished) have confirmed this finding when dried grass rich in sugars was compared with fresh silages. These differences, however, were not apparent when fresh grass was compared with fresh silage and it would be wrong to assume that silage diets necessarily resulted in an unusually high ruminal acetate level. There is little known about the metabolism of silage by ruminal micro-organisms although recently Ewart (1974) has published some data using an *in vitro* continuous fermentation system.

There is very little information available on the net energy value of silages, although in a recent paper by Van der Honing *et al.* (1973) in which silage and hay made from similar grass were compared in dairy cow, energy-balance trials, no significant differences between hay and silage were noted in digestibility and energy or in conversion of digestible energy and metabolizable energy into net

Table 4

The influence of oxygen on some silage characteristics

Treatment	pH	sugars (%)	vol. N (%)	acetate (%)	butyrate (%)	lactate (%)	Total count (× 10⁶)	Lactic acid bacteria (× 10⁶)	Yeasts + moulds (× 10⁴)	Clostridia proteolytic	lactate ferm.
Grass –	6.08	12.49	0.03	–	–	–	176	0.72	> 10	250	30
Silage* 0.26	3.91	0.99	0.27	2.58	Nil	14.42	15.2	25	0.7	1100	< 10
6.59	5.72	1.25	0.96	5.36	3.20	Nil	68.0	132	31	30	250

* O_2/grass (v/v) ratios.
Data from Ruxton & McDonald, 1974

energy. However, the gross energy of the silage organic matter was higher than that of the hay and therefore both metabolizable and net energy content of silage organic matter was *c.* 5–6% higher also. Methane production tended to be slightly higher from silage than the hay diet, although it was suggested that this could have arisen from the action of methanogenic bacteria on the formic acid used as a silage additive.

The chemostatic and physiological mechanisms which control dry matter intake in monogastric animals are considered to apply to ruminants with high digestibility foods such as concentrates (Jones, 1972). With fibrous foods of lower digestibility, the primary determinant controlling intake is the digestion in the rumen. Such diets are digested and metabolized by ruminal micro-organisms at a slower rate and hence, rate of passage or disappearance from the rumen is reduced. Crampton (1957) concluded that voluntary intake of forage was limited primarily by rate of cellulose and hemicellulose digestion. This is dependent upon the degree of lignification in forages of advanced maturity or the amount of nitrogen available to the rumen flora. Silages, however, do not conform to this general pattern and it has been known for some time that intakes of silage dry matter are frequently well below those for fresh or dried herbages of similar digestibility.

Thomas *et al.* (1961) found that the dry matter consumption of silage by ruminants increased with increasing dry matter content and they produced an equation which showed a linear relationship. A similar relationship was also obtained by Jackson & Forbes (1970). Various attempts have been made to explain this relationship, but the depression in dry matter intake when animals are fed wet silages is not caused by moisture *per se.* It has been suggested that appetite-depressants such as aldehydes, histamines and organic acids are more abundant in wet silages than those of high dry matter content (Neumark, Bondi & Volcani, 1964; Harris, Raymond & Wilson, 1966). Studies by Wilkins and co-workers (McLeod, Wilkins & Raymond, 1970; Wilkins *et al.,* 1971; Wilkins & Wilson, 1971) have indicated that silage intake may be limited in different circumstances, either by products of protein degradation in silages which have undergone a clostridial fermentation or by high concentrations of free acids in silages of low pH value. The negative correlation of silage intake with proteolytic clostridial activity, as indicated by volatile nitrogen (NH_3-N) content, has previously been reported by Gordon *et al.* (1961).

McLeod *et al.* (1970) added sodium bicarbonate to low pH (4.1) silages and obtained increases in dry matter intake from 9.7–20.7%. Conversely, the addition of lactic acid to silages decreased voluntary intake. The possible importance of saliva in intake control was suggested by Orth & Kaufmann (1966) who found that saliva secretion was depressed by the infusion of acids into the rumen. However, this would suggest difficulties in maintaining ruminal pH values at normal levels (pH 5.5–6.5). In studies by Donaldson & Edwards (unpublished), the pH value of ruminal contents obtained from fistulated sheep on a range of silage diets did not fall below 6.2 whereas in the same studies, using dried grasses, ruminal pH values fell to 5.7.

The high soluble nitrogen content of silages would suggest that nitrogen utilization by ruminal micro-organisms may be low. Griffiths, Spillane & Bath (1973) have reported that retention of N by ruminants on silage is consistently lower than found on hay produced from the same material. Nitrogen retention is significantly increased by the addition to the diet of carbohydrate (barley) concentrates. Chalmers (1963) demonstrated that ruminal ammonia concentrations were higher in animals fed silage compared with frozen and dried grass. Donaldson & Edwards (unpublished) have shown more recently that the concentration of NH_3 in the rumen was higher in sheep after they were fed fresh silages than after they were fed fresh grass.

6. Fermentation Control

While there is no disagreement about a clostridial fermentation being undesirable, it would seem from the findings reviewed in the foregoing section that some of our original ideas about encouraging an active lactic acid fermentation may need to be revised. Such a fermentation, which results in a low pH product of high acid content, may be acceptable from the viewpoint of fodder preservation but may be undesirable nutritionally, because of the depressing action of such material on voluntary dry matter intake.

In recent years much research has been devoted to a study of techniques which restrict fermentation in the silo. Probably the most direct way of doing this is to wilt the crop in the field to a dry matter content > 28% prior to ensiling. The influence of wilting on composition is shown clearly in Table 5. Very little fermentation occurs in dry silages (haylages), and these materials present storage problems because of difficulties in maintaining anaerobiosis. It is, therefore, necessary to use tower-type silos to prevent excessive oxidation losses. For the more common bunker-type silo a dry matter content of 28–32% in ensiled grass is usual. An alternative method of restricting fermentation is to apply an additive at the time of harvesting. A large number of additives have been examined, and in a recent review of the literature, Mann (pers. comm.) listed 181 different chemical compounds which have been tested over the past 50 years as potential silage additives. It is not possible to review this vast subject in this paper and I propose to discuss only a few additives which are of topical interest.

Additives may be classified broadly into 2 categories, stimulants and inhibitors. Stimulants include molasses and cereal enzyme mixtures which have in the past been applied to sugar-deficient crops such as legumes, but they encourage fermentation rather than restrict it. In the second category, one of the

Table 5

Effect of pre-wilting on silage composition

	Fresh silage	Silage prewilted 29 h	Silage prewilted 52 h
pH	3.7	4.1	4.9
dry matter %	15.9	33.6	46.9
Components of dry matter (%)			
Sugars	1.7	11.7	16.4
Volatile N as % total N	6.9	5.9	4.3
Lactic acid	12.1	5.4	1.7
Acetic acid	3.6	2.1	1.2
Butyric acid	nil	nil	nil
Ethanol	1.0	0.4	0.4
Mannitol	5.6	7.4	3.9

Silages made from first cut of *L. multiflorum.*

earliest inhibitors used was the A.I.V. (A. I. Virtanen) mixture of mineral acids which was popular in Scandinavian countries (Watson & Nash, 1960). In recent years formic acid has largely replaced mineral acids in Scandinavia as well as in other countries. In the U.K., the commercial product in commonest use contains 85% of formic acid and the recommended application rate is 2.27 l (2.72 kg)/tonne of fresh grass. At this level of application, formic acid decreases the pH value of the fresh grass to *c.* 4.6–4.8 but does not inhibit completely the growth of lactic acid bacteria (Henderson & McDonald, 1971). Complete inhibition of lactic acid bacteria requires a concentration 2 to 3 times the recommended level depending upon the type of crop and dry matter content (Henderson & McDonald, 1971; Wilkins & Wilson, 1971). Yeasts are inhibited less by formic acid than are bacteria (Henderson, McDonald & Woolford, 1972). Studies with formic acid have generally indicated that this additive improves the preservation of difficult crops such as legumes and grass low in sugar content. With crops rich in sugars such as *Lolium* spp. which normally preserve satisfactorily, the effects of the additive in improving conservation are more difficult to demonstrate. There are several references in the literature, however, indicating the nutritional superiority of formic acid-treated silages, particularly with respect to improved dry matter intake (e.g. Waldo *et al.*, 1969; Castle & Watson, 1970). The question as to whether formic acid is the best member of the fatty acid homologous series to use as a silage additive is important, especially as there are several reports indicating an increased microbial inhibitory action as one ascends the homologous series (Niemann, 1954; Galbraith *et al.*, 1971; Galbraith & Miller, 1973*a,b*). Galbraith *et al.* (1971) examined fatty acids within the range C_8–C_{18} and concluded that in pure culture studies, Gram positive bacteria were inhibited by the long chain fatty acids in the order $C_8 <$

$C_{10} < C_{12} \geqslant C_{14} > C_{16} \geqslant C_{18}$ and the activity of the C_{18} fatty acids in relation to unsaturation and isomerism was $C_{18:0} < C_{18:1}$ (trans) $< C_{18:1}$ (cis) $< C_{18:2} \leqslant C_{18:3}$. Minimum inhibitory concentrations (mM) of long chain fatty acids against *Bacillus megaterium* were C_{16} (0.3); C_{18} (0.4); $C_{18:1}$ cis (0.05) and $C_{18:2}$ (0.02). The possible mode of action of these acids was discussed by Galbraith & Miller (1973*a,b*). Studies with the higher fatty acids $C_{18:0}, C_{18:1}$ and $C_{18:2}$ as silage additives have been disappointing and when applied at rates of 2.3 g/kg of fresh grass (8–9 mM) using laboratory silos, lactic acid fermentation was not restricted (McDonald & Henderson, 1974). The insolubility of the higher fatty acids presents a major problem in ensuring effective distribution over the large surface area of herbage which is not encountered in *in vitro* culture studies. Of the lower fatty acids, hexanoic acid (C_6) has been shown to restrict lactic acid fermentation at a concentration of 5.8 g/kg of fresh herbage although at lower concentrations the acid appears to stimulate lactate production (McDonald & Henderson, 1974). At present, because of economic factors, formic acid is likely to remain the additive of preference in the fatty acid series.

Recently, attention has concentrated on the use of formaldehyde, and the effects of this sterilant, applied as a 35% (w/w) solution (formalin) have been described by Wilkins, Wilson & Woolford (1973). Application rates of 6.8 l/tonne and above resulted in silages with very low levels of fermentation acids and volatile N. Undoubtedly formalin is an effective sterilant although its use requires care because of the potential effects on the rumen microflora (Ewart, 1974). Wilkins (pers. comm.) has evidence that when formalin is applied to ryegrass at 12.3 l/tonne or over, intake is depressed and rumen function disturbed as indicated by decline in cellulose digestion and volatile fatty acid production in the rumen. Barry, Fennessy & Duncan (1973) fed formaldehyde-treated silages (containing 1.7% (w/w) of HCHO) to sheep and obtained increased dry matter intakes, liveweight gains and wool growth rates compared with those given by untreated silages. Our own studies suggest that sometimes formaldehyde-treated silages may decrease digestible protein values compared with untreated materials. The possibility that formaldehyde/acid mixtures may be more beneficial is at present being examined at Edinburgh and other centres.

7. Aerobic Deterioration of Silage

During the feeding period, silage is exposed for varying amounts of time, to atmospheric oxygen. The stability of silage to oxidation is variable but the available evidence indicates that silages which have undergone a restricted fermentation in the silo are inclined to be less stable than those in which fermentation has been unrestricted.

Aerobic deterioration of silage, in the early stages, is attributable to the activity of yeasts which catabolize fermentation acids and residual sugars

Table 6

Effect of additives on silage deterioration

	Original silages		
	Control	Formic acid*	Formic acid/ formaldehyde†
pH	4.51	4.78	5.00
d.m. %	32.4	32.4	31.4
Bc.‡	70	47	48
Components of d.m. %			
Sugars	5.4	15.6	15.7
NH_3–N	0.12	0.09	0.08
Lactic acid	4.4	0.9	1.9
Formic acid	0.5	1.3	2.1
Acetic acid	2.3	1.2	1.2
Propionic acid	0.03	0.05	0.20
Butyric acid	0.19	0.09	0.30
Ethanol	0.41	0.79	0.75

Microbiological data during deterioration

	Yeast count as % total count		
Time after unloading (h)	Control	Formic acid	Formic acid/ formaldehyde
0	0.3	4.9	2.1
24	0.5	2.6	18.2
48	0.2	31.1	17.7
72	0.2	22.7	40.0
96	–	–	17.0

* Original application of formic acid = 14.0 g/kg grass d.m.
† Original application of formic acid = 10.1 g/kg grass d.m. + formaldehyde = 4.3 g/kg grass d.m.
‡ Buffering capacity, m equiv/100 g d.m.
Mann, Henderson & McDonald – unpublished data

(Zimmer, 1969). Daniel *et al.* (1970) have indicated that silages with $> 10^5$ yeast organisms/g are very unstable and that silages with high acetate and butyrate contents are relatively stable because both these acids are fungistatic. At Edinburgh we are examining the effects of a range of treatments upon the rate of silage deterioration in air using a specially constructed 'Thermalog' apparatus consisting of a 14-channel recording thermistor thermometer which is able to record both absolute and differential temperatures. The apparatus is used to record temperature increases in silages held in insulated polystyrene containers. Figure 1 shows a typical result obtained for an untreated silage, a silage treated with formic acid (14 g/kg dry grass) and one treated with a formic acid (10.1 g/kg) and formaldehyde (4.3 g/kg) mixture (analysis of silages are shown in Table 6).

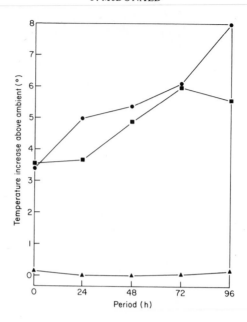

Fig. 1. Effect of additives on silage deterioration. ●, silage treated with formic acid; ■, silage treated with formic acid/formaldehyde mixture; ▲, untreated control.

It seems clear from these results that silages in which fermentation has been restricted are particularly liable to deteriorate rapidly on exposure to air.

8. Conclusions

Our understanding of the ensilage process is sufficiently advanced to be able to control the fermentation by encouraging a lactic acid fermentation and inhibiting clostridial growth. In view of the animal intake problems, however, associated with low pH silages, some restriction in the production of fermentation acids may be desirable. The most direct way of achieving this is to wilt the crop prior to ensiling to a dry matter level of 28–30%. Alternatively an additive may be applied. Research is continuing on the most suitable additive, or mixture, to use. The objectives in the selection of such a mixture are that it should (a) prevent clostridial growth, (b) restrict the activities of the lactic acid bacteria and (c) inhibit deterioration of silage on exposure to air during feeding. The range of additives which can be used in practice is necessarily limited because of the need to avoid any adverse effects on subsequent fermentation by ruminal micro-organisms or on animal performance.

9. References

ANDERSON, B. K. & JACKSON, N. (1969). Conservation of heavily-wilted herbage at four stages of growth in experimental air-tight silos. *Rec. agric. Res. (N.Ir.)* **18**, 95.

ANDERSON, B. K. & JACKSON, N. (1971). Volatile fatty acids in the rumen of sheep fed grass, unwilted and wilted silage, and barn-dried hay. *J. agric. Sci.* **77**, 483.

BARRY, T. N., FENNESSY, P. F. & DUNCAN, S. J. (1973). Effect of formaldehyde treatment on the chemical composition and nutritive value of silage. *N.Z. J. agric. Res.* **16**, 64.

BLAXTER, K. L. (1967). *The Energy Metabolism of Ruminants.* London: Hutchinson.

BROWN, W. O. & KERR, J. A. M. (1965). Losses in the conservation of heavily-wilted herbage sealed in polythene film in lined trench silos. *J. Br. Grassld. Soc.* **20**, 227.

CASTLE, M. E. & WATSON, J. M. (1970). Silage and milk production, a comparison between wilted and unwilted silages made with and without formic acid. *J. Br. Grassld. Soc.* **25**, 278.

CHALMERS, M. I. (1963). The significance of the digestion of protein within the rumen on the nutrition of the animal. In *Progress in Nutrition and Allied Sciences.* Ed. D. P. Cuthbertson. Edinburgh: Oliver & Boyd.

CRAMPTON, E. W. (1957). Interrelations between digestible nutrient and energy content, voluntary dry matter intake and the overall feeding value of forages. *J. anim. Sci.* **16**, 546.

DANIEL, P., HONIG, H., WEISE, F. & ZIMMER, E. (1970). The action of propionic acid in the ensilage of green fodder. *Wirtschaftseigene Futter* **16**, 239.

DEWAR, W. A., McDONALD, P. & WHITTENBURY, R. (1963). The hydrolysis of grass hemicelluloses during ensilage. *J. Sci. Fd Agric.* **14**, 411.

EWART, J. M. (1974). *In vitro* rumen systems. *Proc. Br. J. Nutr.* **33**, 125.

GALBRAITH, H., MILLER, T. B., PATON, A. M. & THOMPSON, J. K. (1971). Antibacterial activity of long chain fatty acids and the reversal with calcium, magnesium, ergocalciferol and cholesterol. *J. appl. Bact.* **34**, 803.

GALBRAITH, H. & MILLER, T. B. (1973a). Effect of long chain fatty acids on bacterial respiration and amino acid uptake. *J. appl. Bact.* **36**, 659.

GALBRAITH, H. & MILLER, T. B. (1973b). Physico-chemical effects of long chain fatty acids on bacterial cells and their protoplasts. *J. appl. Bact.* **36**, 647.

GIBSON, T. (1965). Clostridia in silage. *J. appl. Bact.* **28**, 56.

GIBSON, T., STIRLING, A. C., KEDDIE, R. M. & ROSENBERGER, R. F. (1958). Bacteriological changes in silage made at controlled temperatures. *J. gen. Microbiol.* **19**, 112.

GIBSON, T., STIRLING, A. C., KEDDIE, R. M. & ROSENBERGER, R. F. (1961). Bacteriological changes in silage as affected by laceration of the fresh grass. *J. appl. Bact.* **24**, 60.

GORDON, C. H., DERBYSHIRE, J. C., WISEMAN, H. G., KANE, E. A. & MELIN, C. G. (1961). Preservation and feeding value of alfalfa stored as hay, haylage and direct-cut silage. *J. Dairy Sci.* **44**, 1299.

GRIFFITHS, T. N., SPILLANE, T. A. & BATH, I. H. (1973). Studies on the nutritive value of silage with particular reference to the effects of energy and nitrogen supplementation in growing heifers. *J. agric. Sci. Camb.* **80**, 75.

HARRIS, C. E., RAYMOND, W. F. & WILSON, R. F. (1966). The voluntary intake of silage. *Proc. 6th int. Grassld. Congr. Helsinki,* 564.

HENDERSON, A. R. & McDONALD, P. (1971). Effect of formic acid on the fermentation of grass of low dry matter content. *J. Sci. Fd Agric.* **22**, 157.

HENDERSON, A. R., McDONALD, P. & WOOLFORD, M. K. (1972). Chemical changes and losses during the ensilage of wilted grass treated with formic acid. *J. Sci. Fd Agric.* **23**, 1079.

JACKSON, N. & FORBES, T. J. (1970). The voluntary intake by cattle of four silages differing in dry matter content. *Anim. Prod.* **12**, 591.

JONES, G. M. (1972). Chemical factors and their relation to feed intake regulation in ruminants. A review. *Can. J. anim. Sci.* **52**, 207.

KROULIK, J. T., BURKEY, L. A. & WISEMAN, H. G. (1955). The microbial populations of the green plant and of cut forage prior to ensiling. *J. Dairy Sci.* **38**, 256.

LANGSTON, C. W. & BOUMA, C. (1960). A study of the micro-organisms from grass silage. II The lactobacilli. *Appl. Microbiol.* **8**, 223.

McDONALD, P. & HENDERSON, A. R. (1974). The use of fatty acids as grass silage additives. *J. Sci. Fd Agric.* **25**, 791.

McDONALD, P., HENDERSON, A. R. & RALTON, I. (1973). Energy changes during ensilage. *J. Sci. Fd Agric.* **24**, 827.

McDONALD, P., STIRLING, A. C., HENDERSON, A. R. & WHITTENBURY, R. (1962). Fermentation studies on wet herbage. *J. Sci. Fd Agric.* **13**, 581.

McDONALD, P. & WHITTENBURY, R. (1973). Ensilage. In *The Chemistry and Biochemistry of Herbage.* Eds G. W. Butler & R. W. Bailey. London: Academic Press.

McLEOD, D. S., WILKINS, R. J. & RAYMOND, W. F. (1970). The voluntary intake by sheep and cattle of silages differing in free-acid content. *J. agric. Sci. Camb.* **75**, 311.

MacPHERSON, H. T. & SLATER, J. S. (1959). γ-amino-n-butyric, aspartic, glutamic and pyrrolidone carboxylic acid; their determination and occurrence in grass during conservation. *Biochem. J.* **71**, 654.

MacPHERSON, H. T. & VIOLANTE, P. (1961). The influence of pH on the metabolism of arginine and lysine in silage. *J. Sci. Fd Agric.* **17**, 128.

NEUMARK, H., BONDI, A. & VOLCANI, R. (1964). Amines, aldehydes and keto-acids in silage and their effect on food intake by ruminants. *J. Sci. Fd Agric.* **15**, 487.

NIEMANN, C. (1954). Influence of trace amounts of fatty acids on the growth of micro-organisms. *Bact. Rev.* **18**, 147.

OHYAMA, Y. (1971). Problems in silage fermentation. *Jap. J. Zootech. Sci.* **42**, 301.

OHYAMA, Y., MASAKE, S., TAKIGAWA, A. & MORICHI, T. (1970). Studies on various factors affecting silage fermentation. 8. Changes in microflora and organic acid composition during ensilage as affected by the introduction of air after ensiling. *Jap. J. Zootech. Sci.* **41**, 625.

ORTH, A. & KAUFMANN, W. (1966). The effect of bicarbonate on the feed intake in dairy cows. *Z. Tienerphysiol. Tierernähr. Futtermittelk.* **21**, 350.

PLAYNE, M. J. & McDONALD, P. (1966). The buffering constituents of herbage and of silage. *J. Sci. Fd Agric.* **17**, 264.

RUXTON, I. B. & McDONALD, P. (1974). Influence of oxygen on ensilage. 1. Laboratory studies. *J. Sci. Fd Agric.* **25**, 107.

STIRLING, A. C. (1953). Lactobacilli and silage-making. *Proc. Soc. appl. Bact.* **16**, 27.

STIRLING, A. C. & WHITTENBURY, R. (1963). Sources of lactic acid bacteria occurring in silage. *J. appl. Bact.* **26**, 86.

TAKAHASHI, M. (1970). Influence of level of initial air inclusion in ensiling on quality of silage. *J. Jap. Grassld. Sci.* **16**, 98.

THOMAS, J. W., MOORE, L. A., OKAMOTO, M. & SYKES, J. F. (1961). A study of factors affecting rate of intake of heifers fed silage. *J. Dairy Sci.* **44**, 147.

VAN DER HONING, Y., VAN ES, A. J. H., NIJKAMP, H. J. & TERLUIN, R. (1973). Net-energy content of Dutch and Norwegian hay and silage in dairy cattle rations. *Z. Tierphysiol. Tierernähr. Futtermittelk.* **31**, 149.

WALDO, D. R., SMITH, L. W., MILLER, R. W. & MOORE, L. A. (1969). Growth, intake and digestibility from formic acid silage versus hay. *J. Dairy Sci.* **52**, 1609.

WATSON, S. J. & NASH, M. J. (1960). *Conservation of Grass and Forage Crops.* Edinburgh: Oliver & Boyd.

WEIRINGA, G. W. (1958). The effect of wilting on butyric acid fermentation in silage. *Neth. J. agric. Sci.* **6**, 204.

WEISE, F. (1968). The influence of chopping on the course of fermentation of moist silages. *Das wirtschaftseigene Futter,* **14**, 294.

WHITTENBURY, R. (1968). Microbiology of grass silage. *Process Biochem.* **3**, 27.

WHITTENBURY, R., McDONALD, P. & BRYAN-JONES, D. G. (1967). A short review of some biochemical and microbiological aspects of ensilage. *J. Sci. Fd Agric.* **18,** 442.

WILKINS, R. J. (1974). The nutritive value of silages. *8th Nutr. Conf. Feed Manuf. Univ. of Nottingham.*

WILKINS, R. J., HUTCHINSON, K. J., WILSON, R. F. & HARRIS, C. E. (1971). The voluntary intake of silage by sheep. 1. Interrelationships between silage composition and intake. *J. agric. Sci. Camb.* **77,** 531.

WILKINS, R. J. & WILSON, R. F. (1971). Silage fermentation and feed value. *J. Brit. Grassld. Soc.* **26,** 108.

WILKINS, R. J., WILSON, R. F. & WOOLFORD, M. K. (1973). The effects of formaldehyde on the silage fermentation. *Proc. 5th Gen. Meeting of Europ. Grassld. Fed., Uppsala.*

WOOLFORD, M. K. (1972). Some aspects of the microbiology and biochemistry of silage making. *Herb. Abstr.* **42,** 105.

ZIMMER, E. (1967). The influence of prewilting on nutrient losses, particularly on the formation of fermentation gas. *TagBer. dt. Akad. LandwWiss. Berl.* Nr. 92, 37.

ZIMMER, E. (1969). Biochemical principles of ensiling. *Proc. 3rd Cong. Europ. Grassld. Fedn. 113.*

ZIMMER, E. (1971). Fodder conservation as an agronomical and biotechnical problem. *Z. Acher-und Pflanzenbau,* **133,** 85.

Trends and Innovations in Rumen Microbiology

Rowett Research Institute, Bucksburn,
Aberdeen, Scotland

RUMENS have been in existence for thousands of years, perhaps for millions of years as the herbivorous dinosaurs may have been ruminants. Microbiology has existed for more than a hundred years; rumen microbiology has attained its majority only recently. But in 'celebrating' its majority we must ask what rumen microbiology has achieved in its 20 odd years, and what is its future.

The rumen investigations have perhaps progressed farther than those of many other systems, so consideration of the rumen should help us to investigate other microbial ecosystems, partly by indicating unprofitable lines of work.

Now I think that microbiology is probably one of the most 'applied' sciences, in the old sense of 'pure' and 'applied', that there is. It had its beginnings in wondering why the beer went sour, why the dog had rabies, or what was 'Will-o-the Wisp'? We now know more of what microbes look like, what they are made of, how they work. But nearly all this knowledge can be, and is, applied to answering the kind of questions posed originally; can we prevent disease in man, can we stop the machine-tool oil 'going off', can we improve the yield of amino acid, can we alter our microbe's environment to provide man with more fats, proteins or polysaccharides?

In many of these questions we have a relatively straightforward micro-biological problem. We have a microbe, we have a product that we can sell and we have to optimize production of this product on as large a scale as necessary. This is far from easy, but at least we have a definite goal, and by due attention to bio-engineering principles we can isolate our microbe in the system so that it can live on its own and do its job under the most favourable conditions.

In a rumen, sewage works, river or ocean we have a much more diffuse system, exposed to all contamination, and we can have many goals in our investigations. But we have to define our goal and to pick a path for our investigation using as aids all the available microbiological techniques.

Our goal may be multiple, for as the years go by we see new things to aim at. First we have the system, say the rumen, and we must enquire whether it has microbes and what they do. In these investigations we often progress further as new techniques are developed in other fields. But new questions arise such as: do the microbes always do the same thing? Are they always the same microbes? Why do they change? And finally, as I have just suggested that nearly all microbial problems are of interest to man's welfare, can we make the microbes change to order? Can we make them work more efficiently, or in some cases can we prevent them working, to improve the life of man?

All this may appear obvious, but it is so easy to lose one's way. A new technique is developed or a new idea arrives. How nice to apply this to one's own field, but when we have determined the 57th variety of x in the 157th variety of bacterium from the 77th habitat are we any nearer explaining how those seventy seven habitats exist? We must beware of the old adage of knowing more and more about less and less. And I think that to some extent the microbial ecologist, because that is who we are really talking about today, must be a non-specialist, a jack-of-all-trades; and because he is a non-specialist his work may have repercussions beyond his own field.

So let us consider the history and, to some extent, the future of rumen microbiology in this light, an investigation on its own, but taking in from, and giving out information to, other investigations.

The rumen and the complex stomach system of the cow must have been known for hundreds of years but micro-organisms were reported to inhabit the rumen only about 150 years ago. This started an observational phase of rumen microbiology. Observation, as a means of trying to determine what the rumen microbes did, lasted until c. 1950. That is not to say that other things were not being done, but microscopy was used as an investigational tool on its own. But while light microscopy can tell us about the structure of tissues and so on, it can tell us little, if anything, about the biochemical function of cells. Some deductions were made; for instance it was seen that plant fibres harboured bacteria. The fact that a bacterium is on a piece of grass may be purely fortuitous, but if this bacterium is in a hole in the grass then the deduction may be made, and was made, that the bacterium is digesting the cellulose in some way. A protozoon is seen with a piece of grass inside it, and this morphological type of protozoon always seems to have grass in it, but another type seems always to have starch granules in it. One might reasonably suggest that the one protozoon lives on grass and the other on starch. So we can suggest that the microbes of the rumen live on the cow's feed. We might assume that in some way this is necessary to the animal's digestive processes, but observation cannot tell us in what way, and indeed it may mislead us. For instance, staining with iodine showed that many of the bacteria on the grass fibres stained in a manner suggesting that a starch-type polysaccharide was present. So it was suggested that the bacteria converted the indigestible cellulose into starch which could be digested by the cow when the bacteria passed on from the rumen to the lower digestive tract, because there the bacteria were found in various stages of disintegration. We now know that although this digestion of microbial starch takes place to a very small extent, it is not the way in which the microbes help the ruminant to gain energy from plant material.

The true means whereby microbes aid the animals' carbohydrate digestion was found from the work of the physiologist and biochemist who showed that in the rumen, and in other parts of the intestines of ruminants and non-ruminants, where large numbers of bacteria were found, there were also relatively high concentrations of the 3 lower volatile fatty acids, acetic, propionic and butyric. These acids were shown eventually to be the substances used by the ruminant for production of energy and tissue constituents such as fats. The bodies of the micro-organisms provide a source of protein for the animal.

The microscopist was still recording the shapes of the bacteria and protozoa, but other microbiologists were thinking of further steps, and biochemical and physiological work on rumen function was continuing. It was known that the occurrence of volatile fatty acids in the digestive tract was accompanied by anaerobic conditions and it seemed that the acids were produced by microbial fermentation. It is possible to work on the biochemistry of the rumen without ever actually isolating a bacterium, either by examining the contents by sampling from the rumen or by taking out a portion of the contents and incubating it *in vitro*. Both these methods have told us much about the overall rumen function. but they have limitations, and many of these limitations have only been shown by microbiological work. The *in vivo* method can show the overall effects of a change in diet of the animal, but because of the complexity of the rumen contents it is difficult to study the fate of one dietary component, and because of the size of the rumen one is generally limited to studying the effects of major substrates for the microbes; it is difficult to follow a small amount of some intermediate in a reaction pathway and it is difficult, or impossible to change only one parameter of the system. The *in vitro* method has a number of disadvantages. Reactions generally have to be short term or end products build up; if washed suspensions are used, the micro-organisms are not growing; if whole rumen contents are used, microbial products and feed-stuff particles are present initially and the organisms may or may not be growing. The micro-organisms are subject to a change in environment on removal from the rumen and even if this does not lead to gross observable changes there may be effects on particular constituents of the complex microbial population. Attempts have been made to expand the simple 'flask' *in vitro* system into an artificial rumen based on continuous culture, as we shall see later, but these again have limitations.

The methods using rumen contents and the mixed populations of the rumen can show the overall effects of adding substrates, growth factors, inhibitors, and so on, on the reactions of the rumen micro-organisms. But to be at all effective they must be subject to strict bacteriological control and even then there are limits to what can be done and what can be deduced. To gain a deeper insight into the microbial activities in the rumen, pure cultures, or defined mixtures must be studied under defined conditions *in vitro*. And this means the micro-organisms must be isolated.

Of course, much of the work we have just briefly discussed, and especially that with the artificial rumen, is part of comparatively recent research and has continued in company with the cultural work, but a few workers attempted the isolation of rumen bacteria long before the intensive work on rumen function of the last 20 or 30 years began.

Earlier workers using what were then conventional media and methods obtained 1 or 2 cultures of bacteria from the rumen. Some workers attempted to culture cellulolytic bacteria. But the important rumen bacteria could not be cultured. In the late 1940s 2 or 3 workers first isolated true rumen bacteria.

Gall and her co-workers realizing the anaerobic nature of the rumen used anaerobic media, but made up from the usual bacteriological protein hydrolysates, etc., to give a rich broth. Bacteria morphologically like some of those in the rumen were isolated, and in high numbers. It is doubtful if cellulolytic bacteria were isolated. At about the same time 2 other workers were concentrating on isolating cellulolytic bacteria, as breakdown of cellulose in plants is one of the major reactions of the rumen. Sijpesteijn in 1948 isolated a cellulolytic coccus which subsequent work has shown to be one of the most important rumen cellulolytic bacteria. Although both workers used reduced media, Hungate (1947) rather earlier had used the results of the biochemist and devised a medium which better simulated the chemical habitat of the bacteria. This of course was not an entirely new idea, but it was perhaps a new idea from the point of view of microbiologists dealing with bacteria in living bodies, as although some had used the dilute media of the soil bacteriologist, other workers had tended to follow the medical microbiologist and use complex and rich media made from infusions of tissues. This may simulate the habitat for bacteria growing in a wound, but chemical analysis shows the rumen liquid to be very low in amino acids and free sugars, and high only in ammonium. So Hungate's medium contained ammonium as principal nitrogen source and some rumen fluid as a further source of growth factors, a salt mixture similar to rumen fluid and a bicarbonate buffer with an atmosphere of carbon dioxide, the principal gas in the rumen atmosphere. He also devised a method for preparing media and cultures such that highly reduced conditions could be maintained. These media and methods have remained the basis of rumen microbiological techniques, and, it might be noted in passing that the same methods enabled Hungate to isolate a cellulolytic bacterium from an anaerobic sewage digester.

The methods resulted in the isolation of cellulolytic bacteria that produced some of the acids found in the rumen contents and gave proof of the connection of the bacteria, the plant fibres and the acids which the animal used. But the rumen fluid seemed an essential constituent for growth in pure cultures of some of the rumen bacteria and Sijpesteijn found that, in her medium, her coccus seemed to grow well in conjunction with a clostridium or with contaminating bacteria from the dilutions of rumen contents; the cocci needed something produced by the clostridium, that was not in the medium constituents. The nature of this substance could not be determined. These early results then suggested the important and fascinating problem of the interdependence of the rumen microbes, a problem that is still being looked at today.

Following these early experiments there began a period when many rumen bacteria were isolated and described, and to some extent this continues today. Although rumen microbiology has never attracted large numbers of workers much work has been done, and many bacteria isolated.

There are 2 ways of isolating bacteria from a habitat such as the rumen. One is to use non-specific media and as many media as possible in the hope of isolating all the bacteria. The reactions of these can then be tested *in vitro* and they can be classified. From this one can get a description of the bacterial population, but unless the *in vitro* tests are selected properly one can get little idea of the role that these bacteria play in the habitat. The other method is, as in the earlier experiments mentioned, to use a medium that one hopes will be selective for bacteria performing a specific biochemical action known to be important in the habitat. The bacteria growing in this medium can then be examined and classified.

The second method, of course, is only of value if at least the overall biochemistry of the habitat is known. In the case of the rumen we have the situation where far more people have worked on the biochemistry than on the microbiology, of the organ. So to some extent the microbiologist has always been in the position of finding a bacteriological explanation for a chemical reaction. This is probably the best way, or the only way, of investigating such a microbial habitat in the beginning, but at some point the microbiologist must begin to take the lead.

However, the period of isolation of bacteria was accompanied by investigation of their biochemistry. This showed the substances fermented and the fermentation products, and the nitrogen requirements and growth factors needed. The first showed that the bacteria almost without exception obtained energy from fermentation of carbohydrate and that they produced a mixture of acids and gases as fermentation products. Although these acids numbered amongst them the 3 lower volatile fatty acids found in the rumen this type of work does little to explain why virtually only these 3 acids are found and why they are in certain proportions. However, the work on nitrogen sources and growth factors was more productive in showing the way in which the rumen microbes exist as a symbiotic population. It showed that there were proteolytic bacteria that hydrolysed the feed proteins to amino acids and peptides that some bacteria could use as nitrogen sources, but that many bacteria needed ammonia alone or ammonia plus amino acids, and that this ammonia was produced by certain of the bacteria which deaminated the amino acids. Rumen fluid, added to the earliest media, had been found necessary for growth of a number of the bacteria and the factors in this were shown to be the volatile fatty acids. Not so much the 3 main ones, although acetic acid is a growth factor for some rumen bacteria, but the small amounts of higher straight- and branched-chain acids produced by the deamination of amino acids just mentioned.

Bacteria producing methane from hydrogen and carbon dioxide or formate were also isolated and tested and so another step in the rumen fermentation pathway was disclosed.

Isolation studies showed that the same, or similar bacteria existed in rumens all over the world, in the same, or different species of animals. The predominant types varied, more generally with feed than with animal species, but no bacteria appeared to be peculiar to a particular animal.

It was suggested earlier, although not described in detail, that the success of the rumen microbiologist has been due to the adoption of particular techniques for isolating and studying the anaerobic bacteria. The application of these techniques has resulted, in the last few years, in the finding of rumen-type anaerobic bacteria in other habitats such as the intestines of ruminants and non-ruminants, including man; pathological lesions, and places outside the body such as muds and sewage digesters. Rumen microbiology is thus beginning to lead to advances outside its original field.

I have not so far mentioned the rumen ciliate protozoa. These organisms, like the bacteria, exist in other anaerobic habitats in the gut of herbivores. The species may differ, but so far as we know they play a similar role in each habitat. Their biochemistry is similar to that of the bacteria in that they obtain energy by carbohydrate fermentation. However, it was early recognized by observation that they also fed on bacteria and later work has shown that they can use the protein and nucleic acids of the bacteria as sources of amino acids and nucleic acid fragments, respectively, from which to build their own macromolecules.

Early attempts to culture the protozoa showed that they could be kept alive only in very mixed media containing not only bacteria but also materials like starch and grass fibres, and the media had to be changed at regular intervals even when comparatively small populations of protozoa were present. The culture of the rumen protozoa has really not progressed much beyond this stage and all work on the biochemistry of the protozoa has been done on washed suspensions obtained either from the rumen directly or from cultures. The ruminant itself can be used as a culture vessel for the protozoa, in that because of their large size and the difficulty of airborne transfer of the organisms, sheep or cattle may be kept for long periods either free of protozoa or inoculated with specific protozoa. This fact allows a number of deductions to be made about the protozoa, and about their interactions with the other microbes, but the complexity of the system necessarily limits what can be done. Proper continuous culture of the protozoa would allow more to be done, as it is for the protozoa of activated sludge, but no one has yet succeeded in creating such a system.

So far we have dealt with the beginnings of rumen microbiology and with some of its later ramifications, and we saw that this had its beginnings in culture of the micro-organisms in the classical way in tubes. Now we must begin to consider other aspects of rumen microbiology.

So let us look at the classical culture of rumen bacteria. Culture and some form of classification of the bacteria was obviously a primary step in investigation of the rumen, but is it still necessary? The majority of rumen bacterial types have probably been cultured, and we know their properties, as these are defined in the ordinary small, batch culture. The word 'types' is used deliberately because 'species' are a matter for the taxonomist. It was early recognized that there were many strains of rumen bacteria which were similar in major aspects, but differed in a few properties (this, of course, is not confined to the rumen). So should these be put in the same species, or different species or varieties? The taxonomist attempts to answer these questions by comparing as many cultures as possible in as many ways as possible and hopes to produce order out of chaos. Whether it is worthwhile trying to make anything more than a general classification of bacteria is outside the scope of this paper, but I venture to state that the continued isolation and classification of rumen bacteria, by the same methods, from more animals will, in general not lead to a better understanding of rumen function. So let us look at this in more detail.

Representatives of most or all of the species of rumen bacteria have probably been isolated and their properties determined in classical batch culture. Examination of rumen contents is usually done by enumerating dilution cultures on one of the media designed to culture, non-selectively, all the bacteria. Because of the time involved in testing, most surveys of the bacteria in rumens can only indicate that on the basis of a few tests the bacteria isolated are similar in general properties to species already named. This should then tell us the species of bacteria that form the predominant flora of that particular rumen, assuming that our medium will grow all the bacteria. But technical difficulties of examining enough colonies generally mean that we can identify only a certain proportion of the bacteria growing in the highest dilutions. Thus we may be able to say, if there are predominant types, that 50% of the population is of a certain species. What is more difficult to say is what constitutes 5% of the population. We can, then, say that the population of a rumen consists of approximately so many percent of this and that bacterium, and we may be able to say that this appears to be characteristic of, for instance, the rumens of sheep fed on hay. But this is purely descriptive and it tells us nothing about why this population is present and it may not tell us the most important bacteria in this population, because not only are the numbers of culturable or observable bacteria in a rumen at a given time not necessarily connected with the rates of metabolism of these bacteria over a prolonged period, but mere numbers are not a guide to the importance of bacteria *in the functioning* of a microbial system. Let us take another example. In anaerobic digesters fermenting piggery waste, streptococci form one of the largest components of the population but in domestic sewage digesters coliforms form a large percentage of the population. But the presence of these bacteria is due solely to massive inoculation from the incoming sewage, those digesting the waste and controlling the reactions are all of other species. A similar situation is less likely to occur in the rumen because the inoculation from the air and feedstuffs is usually relatively small, but cases have been cited of bacteria present in large numbers only because of inoculation from the feed.

Assuming the main organisms we have identified are truly growing in the rumen, then they may still not be the important ones because their growth may be controlled by products of other bacteria. It is common in a system like the rumen to find that the bacteria responsible for the primary hydrolytic actions such as cellulolysis and proteolysis are only a small proportion of the population; they are in the 5% mentioned previously. In most cases these bacteria control not only the primary supply of substrates for the other bacteria, but, by controlling the rate of breakdown of the host animal's feed, control the whole rumen function. So this means that either we must isolate many bacteria from our non-selective medium and test every isolate for the particular hydrolytic activity, or resort to media that are more selective in order to detect the bacteria causing the particular hydrolysis. Where no visible action on media can be observed we may have to resort to a medium that encourages growth of bacteria with the particular action, and then, (because no medium is truly selective) test a large number of isolates for the action. Each isolate must be tested because some hydrolytic properties are functions of some strains of many species, and although some properties are confined to certain genera they are found in only some strains. So while the initial broad classification may show that a bacterium can be grouped with the species *A.b.* it will not tell us if it is a cellulolytic strain of this species. But if we identify the cellulolytic bacteria we shall need more information because they vary in rate of cellulolysis and ability to hydrolyse different celluloses, so one strain may have overall control of rate of the hydrolysis.

However, we can then go further because, to continue with our example of cellulolysis, some of the cellulolytic bacteria need ammonia, or amino acids plus ammonia as nitrogen sources. The rate of supply of these will depend on the activities of the proteolytic and deaminative bacteria and these again may be only a small fraction of the population. The deaminative bacteria provide the branched-chain volatile fatty acids necessary for some cellulolytic bacteria and provision of these may be rate-controlling. As a further example, some of the cellulolytic and other bacteria produce hydrogen; removal of this hydrogen by the methanogenic bacteria can control the fermentation and growth of the hydrogen-producers. Methanogenic bacteria are seldom, if ever, considered in surveys of rumen bacteria, largely for technical reasons.

There is also the difficulty of bacteria attached to feed particles; these are almost impossible to remove in a viable form of culture. Determination of fermentation rates of rumen solids and liquids suggests that there are at least as many, and possibly more, bacteria attached to feed particles than are free in the liquid. It is true, as might be expected, that representatives of all rumen bacteria are found in the liquid, but it is possible that the relative numbers of the 'attached' bacteria are underestimated in counts. This only covers some of the main points, but bacteria present in low numbers have other vital roles to play in producing vitamins or other growth factors, or in conditioning the medium in terms of E_h or removal of inhibitory substances, for growth of the main hydrolytic or other bacteria.

Similar considerations apply to the protozoa which can be seen and counted. We know some of their main fermentative and hydrolytic actions, but we do not know if particular 'strains' of the morphologically grouped species are more active than others, as with the bacteria. We know that they ingest bacteria which provide precursors of protein and nucleic acids, but we do not know all the interactions of growing protozoa with the bacteria or what governs the rates of growth and metabolic reactions of the protozoa. They are there or not there but exactly why, or what they are doing and at what rate they are doing it is obscure.

To return to the bacteria, we originally made the proviso that our medium, or media, would grow all the rumen bacteria. We can say that all known species of rumen sugar-fermenting bacteria, say, will grow on a non-selective counting medium, when inoculated in pure culture from a laboratory culture, and that experience suggests that they will grow on this medium when the inoculum comes from the rumen. What is not so certain is whether this medium will grow in correct numbers and proportions all these bacteria from different rumen samples. The proportion of apparently viable bacteria in rumen samples is nearly always very low. This may be due to the fact that most of the bacteria are probably 'dead', and there is evidence that this is so, or it may be due to deficiencies in the media; different media give different counts, although these are usually of the same order. Or it may be due to the fact that we still have not cultured all the rumen bacteria. It is known that some of the morphologically distinct large bacteria have not been cultured, or can be cultured only in specialized media and will not grow in the usual cultures used to enumerate the smaller bacteria. But it is very difficult on morphological grounds, the only method possible, to compare the bacteria in culture with those in the rumen. Another method of comparison, the use of fluorescent antibodies, proved generally impracticable because of the diversity of antigenic types of the different species. If there are still uncultured bacteria then obviously more cultural work using our present media and methods is not going to help; we need a radically different approach.

Cultural and microscopic work, then, can tell us with varying degrees of precision depending on media used, the numbers of cultures examined and in what depth the examination takes place, the principal bacteria and protozoa associated with a rumen sample. If sufficient samples are analysed it can tell us whether a population is relatively stable and always associated with a certain feeding regime. But it can tell us little about the importance of the bacteria in the population, for the kind of reasons we have just been discussing. In certain specialized cases, usually connected with rumen malfunction, the flora may become simplified with only 1 or 2 principal types of bacteria present and this flora may obviously be associated with particular fermentation products. Lactic acidosis associated with a predominantly lactobacillus flora is a well-known case in point. But the observation does not tell us why this predominance of lactobacilli occurs.

Cultural work can tell us that certain bacteria inhabit the rumen and it can provide bacteria for further study, but it is this further study which has shown and will show, more about how the rumen works and perhaps how to increase its efficiency. So let us consider other work in rumen and related microbiology.

Cultural work on the rumen bacteria has been paralleled by analysis of rumen contents and both *in vivo* and *in vitro* work on the extent of digestion of various feedstuffs and feed components and analysis of the overall products of this digestion. All this knowledge must be integrated into the more microbiological studies of the rumen.

As was mentioned before initial cultural studies were followed by determination of the substrates fermented by the bacteria, the fermentation products, and the nitrogen, sulphur and growth factor requirements of the bacteria. The latter investigations showed the importance of ammonia in the nitrogen sources of the principal bacteria, and the findings of the branched-chain volatile fatty acids as growth factors was followed by the determination of the role of these acids as carbon skeletons for amino acid synthesis. The carbon dioxide of the rumen atmosphere was found to be necessary for growth of many of the bacteria and its combination into bacterial cell substances, as for instance, the carboxyl groups of amino acids, or into fermentation products, was demonstrated.

This work using the techniques of the biochemist along with bacteriological techniques showed aspects of the interdependence of the rumen bacteria essential for a complete analysis of the ecology of the system, and also gave information of use in the feeding of animals or in the explanation of some of the findings of the nutritionist. The information could begin to suggest ways in which rumen function might be improved or controlled. For instance in some circumstances of feeding a non-protein diet to ruminants (that is using urea or an ammonium salt as a nitrogen source, or supplementing the nitrogen in a protein-deficient herbage with urea) because of lack of amino acids the branched-chain fatty acids can become a limiting factor. Addition of these latter acids to the diet will improve cellulose digestion by increasing growth of cellulolytic bacteria.

The varied needs of the bacteria for nitrogenous substances can be related to analysis of the rumen and such requirements are not likely to vary much with growth conditions of the bacteria. However, the problem of fermentation products of the bacteria was seen early on. A major fermentation product of Sijpesteijns' cellulolytic coccus was succinic acid, but succinic acid was not found in analyses of rumen contents. Following Johns' (1948) finding that succinic acid was decarboxylated to propionate by bacterial fermentation, Elsden & Sijpesteijn (1950) showed by incubations *in vitro* of rumen contents that succinate was converted to propionate in the rumen. Elsden had earlier (1945) shown that lactate was converted to volatile fatty acids in rumen contents. So here were 2 experiments showing a reason why the varied mixture of fermentation products found in pure cultures of the rumen bacteria is not duplicated in the rumen; some primary fermentation products are substrates for a secondary population of bacteria. The problem of these secondary fermentations has continued to be investigated up to the present. The use of radioactively-labelled succinate extended the earlier observations and showed that the major proportion of rumen propionate could be formed indirectly through succinate, although a number of rumen bacteria produce propionate directly from sugars. What has not been entirely solved is which bacteria produce the propionate from succinate. The *Veillonella gazogenes* that Johns showed to decarboxylate succinate are generally present in rumen contents, but only in low numbers. And using the further technique of attempting to compare the specific activity of the bacteria *in vitro* with the numbers *in vivo* suggested that the numbers of the bacteria in the rumen were not sufficient to decarboxylate succinate at the rate found in rumen contents. Later work, when fermentation pathways in bacteria had been worked out suggested another way in which succinate might be decarboxylated, by incorporation into the fermentation pathways of a bacterium producing propionate from carbohydrate *via* the succinate pathway, but evidence for this obtained by 2 different methods, *in vitro*, and *in vivo* with gnotobiotic lambs, is inconclusive.

The story of succinate metabolism, as outlined here, shows how a number of techniques can be integrated to investigate a problem. Some of these experiments depended on methods not readily available earlier, such as use of radioactive isotopes or gnotobiotic lambs. Others brought in information which had been obtained following investigation of bacteria from other habitats, such as that on fermentation pathways.

The metabolism of lactate has also been investigated by a number of techniques by different workers, but the story is more complicated as lactate can be a precursor of acetate and propionate and the propionate can be formed by 2 different pathways in different bacteria. The proportions of the 2 acids and the relative contributions of the 2 pathways to propionate formation can vary with the diet of the animals and the conditions in the rumen. Some possible explanations for these phenomena can be deduced from continuous culture experiments on rumen and other bacteria.

Another technique which has been used in this kind of work is that of enzyme analysis. In this the activities of various enzymes, in, say, fermentation pathways, are measured, in animals on different feeds, or at different times after feeding.

Previous biochemical work on fermentation pathways of bacteria, both non-rumen and rumen, must of course be used as the criterion for deciding on the enzymes to be assayed. Changes in enzyme activities may be ascribed to growth of bacteria or to the presence of different bacteria in different samples. If a specific enzyme can be linked to a particular bacterium, then the presence or absence of the enzyme might be linked to the presence or absence of the bacterium. However, because of the multiplicity of bacteria in the rumen the results of enzyme assays may be difficult to interpret, and although they may indicate that there are differences between rumen samples they do not tell us how these differences occur. But again the results can be considered with other observations to make a more adequate description of rumen function.

So far, then, we have obtained a picture of the rumen from identification of the bacteria and protozoa there, from examination of these organisms in batch culture, in suspensions, and from enzymic assays and isotopic techniques. We have also found out a number of things from investigation *in vitro* of whole rumen contents or fractions in short-term experiments, and from analysis of the rumen contents *in vivo*. But still more knowledge of microbial physiology is needed to explain many aspects of the behaviour of the rumen population.

The whole rumen system *in vivo* is too complicated for detailed analysis and it is very difficult to alter one parameter. Let us consider pH as an example. Alteration of the pH *in vivo* is only really possible by increasing the easily-fermentable carbohydrate in the diet and so increasing fermentation acids to overcome the buffering action of the saliva. This has thus altered at least 2 parameters, energy source and pH, and the resultant change in pH value may not be stable. Taking out some rumen contents into a flask, altering the pH and then incubating the flask not only has all the drawbacks inherent in the *in vitro* batch incubation, but also subjects the micro-organisms to a sudden pH shock, which can upset their metabolism. Two things may happen in the natural rumen with a pH change. In the short term we may get an alteration in fermentation pathways of the existing bacteria and in the long term we may get a change in flora to one more suited to the altered pH. We can look at this in 2 ways *in vitro*, but both involve continuous culture, and continuous culture is a useful method of looking at the function of a system such as the rumen which is a form of continuous culture, although very complicated, and continuous culture allows us to look at bacteria under closely defined conditions.

The first method of continuous culture is the 'artificial rumen'. In this, rumen contents are cultured in a flask or, to simulate the absorptive capacity of the rumen epithelium, a dialysis vessel, and fed with substrate, either pure compounds or ground-up feedstuffs, and with artificial saliva, while the spent medium and cells overflow. This type of apparatus has been made by a number of people but none has been entirely successful in keeping the balance of microbes, most obviously the protozoa, of the original rumen inoculum. Nevertheless some kinds of apparatus can give at least a good approximation to rumen function in terms of fermentation acid production and feed digestion and it is possible to demonstrate such things as effect of pH, with a constant ration, by a pH-stat adding acid or alkali and so to show the effects on fermentation patterns. And with the better apparatus that can keep running continuously for a few weeks the changes in flora consequent upon a prolonged change in pH can be shown.

However, apart from the drawbacks mentioned the artificial rumen suffers from the disadvantage of the natural rumen in containing a complex mixture of micro-organisms which is difficult to analyse. The other method of using continuous culture is to use pure cultures or defined mixtures. In some respects this is more difficult than using an artificial rumen in that the anaerobic bacteria are, for instance, more susceptible to traces of oxygen when in pure culture than when in the mixed rumen culture. On the other hand the system is simpler to analyse. By selection of the bacteria to be investigated to cover types concerned in different rumen reactions, or representatives of typical rumen species, the effects of different cultural conditions can be used to explain or predict reactions occurring in the natural rumen. For instance the effects of pH on fermentation products or the fact that growth of some bacteria is retarded or stopped at certain pH values, could be used to predict the effects of rumen pH changes just mentioned. The effects of culture conditions on the yields and activities of particular enzymes concerned, for example, in feed hydrolysis could be important in optimizing digestion. Growth yields of bacteria are important in the overall efficiency of the rumen because the ruminant utilizes the microbial cells formed in the rumen as sources of amino acids. Maximum growth yields and the effects of various conditions on yield can best be investigated in continuous culture. Actual microbial growth is very difficult to estimate in the rumen or the artificial rumen because the system contains not only a mixture of microbes, but food particles and animal secretions as well. It was mentioned earlier on that determination of numbers of bacteria in a rumen cannot necessarily give an indication of the biochemical role of the bacteria. The rumen bacteria generally ferment a number of carbohydrates and the rumen contents contain more than one possible substrate for each bacterium. Diauxic growth may take place and as this may vary with strain of bacterium continuous culture experiments can help to define which bacteria may be using what substrates under particular conditions.

From the continuous pure culture this type of work can be extended to defined, mixed cultures and the interactions which play a predominant part in the metabolism of a system such as the rumen can be investigated. Such work is only just beginning, with 2-component cultures, because those with more than 2 components become increasingly difficult to analyse, but already it has yielded valuable information.

However, although cultures *in vitro* can bring to light many facts about rumen function no apparatus can duplicate the animal exactly. Apart from possible unknown animal factors there are the effects of varying rates of saliva flow, rumination, mixing (which is different from mixing *in vitro*), and varying residence times of particles because the rumen outflow is not a purely mechanical overflow but selects for particle size.

At first sight it might seem easy to inoculate the rumen of a young animal with a certain flora and allow this to develop, but even if a ruminant is kept apart from its fellows it will pick up bacteria from the air. However, isolation of the animal can prevent the rumen from becoming inoculated with the ciliate protozoa which cannot be carried far in the air, and so defined rumen populations of protozoa can be made by manual inoculation with specific protozoa. The inoculum protozoa can be picked out under a microscope from other rumen contents. But the development of a very mixed bacterial flora from chance inoculation cannot be prevented. The only way to prevent chance contamination and allow inoculation with a defined flora is to use a gnotobiotic animal which can be inoculated with a defined rumen flora. In this way a flora interacting in ways defined or deduced from experiments *in vitro* can be inoculated and its ability to perform normal rumen functions determined, alternatively different components of a flora can be added at intervals and their functions in modifying fermentation pathways and other aspects of rumen metabolism investigated. However, like any other experimental tool the gnotobiotic ruminant has disadvantages, the apparatus is elaborate and from the experimental point of view there is the need to sterilize everything that enters the isolator.

And so the wheel has turned full circle. We started off some 20 or 25 years ago by attempting to take the rumen to pieces and identify its components. Now we are attempting to put those components together again to make a functioning rumen.

On the way between these 2 ends a large number of experimental techniques have been used. And this has been not just an exercise in microbiology because the microbiology has always been associated with the results of the animal physiologist and nutritionist. There is a limit to what each method can tell us and no method can tell us all, but by combining results we begin to see more clearly how a system functions. Clearly, all knowledge and all methods must be integrated into an ecological study.

To some extent I have simplified this story of rumen microbiology. As in most scientific work the story has not unfolded logically. Experiments have been made and forgotten about, it is often difficult to see the beginnings of ideas, and people have independently come to similar conclusions or made similar advances at the same times. But the amount of knowledge has grown and the time has now come when the microbiologist can at least begin to provide rational explanations of rumen function and can say what will probably be the results of various feeding practices or attempts to control the rumen conditions and suggest ways in which rumen performance might be improved. But there is still much to be investigated. For instance the prevention of methane production is thought desirable when looked at solely from the point of view of methane as a waste of energy, but this prevention can set off a whole series of actions and interactions in the rumen population and these overall effects must be considered. We know the bacteria that are, or could be involved, but this type of work needs investigation of metabolic pathways and the effects of culture conditions and mixed culture growth on these: an extension of the type of experiments discussed in the previous pages.

The rumen is, of course, only a part of the ruminant digestive tract; the true stomach and intestines follow. In the caecum and large intestine another microbial habitat is established where feed residues from the rumen can be fermented. The rumen influences reactions here, not only by virtue of the amounts of feed residues left after rumen fermentation, but because the bacteria from the rumen which escape death in the gastric secretions can colonize the caecum. So although it is technically more difficult to investigate the caecum the methods of rumen microbiology can be applied to this organ.

Since the rumen bacteria are contained in living tissue they or their by-products can influence this tissue. It has been known for a long time that the physical development of the rumen is linked with the development of a fermentative and functioning rumen population and, of course, the influence of the intestinal microbial population on the physical structure of the intestines of non-ruminants is known from experiments on gnotobiotic animals. The rumen and intestinal floras must have other effects on the animal. These areas are all being investigated as part of the ecological analysis of the rumen. They are part of the general problem of investigation of any microbial habitat — the influence of the microbial population of the habitat on its surroundings.

But, finally, we might wonder if having got so much knowledge of the rumen and its microbes we can now do away with the animal entirely? Experimental animals are expensive, so do we need actually to feed animals to find out whether a new diet is suitable or how to change an existing diet to obtain certain results? I said we could now make predictions, but these are mainly qualitative; can we take all our knowledge of microbial behaviour and with data on animal feed and feeding pattern compute quantitatively the resultant rumen fermentation? The behaviour of a simple, defined substrate, with a one-bacterium fermentation can be modelled mathematically with reasonable success. These models can be applied to a complex system such as a rumen only by assuming that it is, in effect, simple; that one reaction or one bacterium is rate-controlling. And then, of course, the model can make only a simplified prediction. For more exact and extensive quantitative predictions we need to incorporate into the model a large amount of data on the behaviour of microbes in pure and mixed culture on dissolved and solid nutrients. That computer simulations can be made from the type of data I have been discussing in this paper, and that they can have some success in predicting the digestion of dietary components or the effects of changing diets, is a measure of the success of the microbiologist in sorting out the microbes and the reactions that go to make up a rumen population. That the computer simulations are limited shows that we still have much to learn about microbes and their interactions in complex environments.

This paper gives a general outline of work on rumen microbiology without detailed descriptions of experiments. It seemed that no detailed bibliography was necessary, and in fact it would be impossible to refer to all papers on the different aspects of the subject. However, a few pioneering experiments are mentioned and references to them are given.

References

ELSDEN, S. R. (1945). Volatile fatty acids in the rumen of the sheep. *Proc. Nutr. Soc.* **3,** 243.

ELSDEN, S. R. & SIJPESTEIJN, A. K. (1950). The decarboxylation of succinic acid by washed suspensions of rumen bacteria. *J. gen. Microbiol.* **4,** xi.

GALL, L. S., STARK, C. N. & LOOSLI, J. K. (1947). The isolation and preliminary study of some physiological characteristics of the predominating flora from the rumen of cattle and sheep. *J. Dairy Sci.* **30,** 891.

HUNGATE, R. E. (1947). Studies on cellulose fermentation. III, The study and culture of cellulose-decomposing bacteria from the rumen of cattle. *J. Bact.* **53,** 631.

JOHNS, A. T. (1948). The production of propionic acid by decarboxylation of succinate in bacterial fermentation. *Biochem. J.* **42,** ii.

SIJPESTEIJN, A. K. (1948). Cellulose decomposing bacteria from the rumen of cattle. Ph.D. Thesis, Leiden.

General References to Rumen Microbiology

ALLISON, M. J. (1969). Biosynthesis of amino acids by ruminal microorganisms. *J. Anim. Sci.* **29,** 797.

BALDWIN, R. L., LUCAS, H. L. & CABRERA, R. (1970). Energetic relationships in the formation and utilization of fermentation end-products. In *Physiology of Digestion and Metabolism in the Ruminant*. Ed. A. T. Phillipson. Newcastle upon Tyne: Oriel Press.

BRYANT, M. P. (1959). Bacterial species of the rumen. *Bact. Rev.* **23,** 125.

HOBSON, P. N. (1971) Rumen microorganisms. *Prog. ind. Microbiol.* **9,** 41.

HOBSON, P. N. (1972). Physiological characteristics of rumen microbes and relation to diet and fermentation patterns. *Proc. Nutr. Soc.* **31,** 135.

HUNGATE, R. E. (1966). *The Rumen and its Microbes*. New York & London: Academic Press.

Microbial Interactions in Foods:
Meats, Poultry and Dairy Products

A. A. KRAFT, J. L. OBLINGER*, H. W. WALKER, M. C. KAWAL,
N. J. MOON AND G. W. REINBOLD †

*Department of Food Technology, Iowa State University,
Ames, Iowa 50010, U.S.A.*

CONTENTS

1. Introduction

INTERACTIONS among various micro-organisms in foods may help to explain why certain species of a mixed population become dominant over others. Most food fermentations rely on such results of microbial associations, but competitive situations also exist among microbial flora of all foods, whether by design or by chance. Growth of bacteria in mixed cultures was reviewed recently by Meers (1973), who described 6 different types of microbial interactions: 'competition, commensalism, mutualism, amensalism, predation, and parasitism'. In many instances, the distinctions may not be well-defined, but the conditions existing in foods fall into the common broad categories of either helpful or harmful relationship for the organisms concerned. It is not the purpose of this paper to dwell on technicalities of the relationships, but rather to review observations of associations of some organisms of importance in dairy products, meats and poultry.

Compatibility or dominance of lactic starter cultures growing in association has formed the basis for several investigations which were reviewed by Reddy, Vedamuthu, Washam & Reinbold (1971). These workers used a differential medium for determining strain interactions of mixtures of *Streptococcus lactis* and *Strep. cremoris,* and later worked with mixtures of 3 strains, including *Strep. diacetilactis* (Reddy *et al.,* 1972). Combinations of the organisms revealed

* Present address: Department of Food Science, University of Florida, Gainsville Florida 32601, U.S.A.
† Present address: Leprino Cheese Company, P.O. Box 8400, Denver, Colorado 80201, U.S.A.

141

definite dominance or compatibility among the strains; from a practical standpoint, the investigators emphasized that compatibility would be desirable for commercial application.

Antagonistic action of enterococci (*Strep. faecalis, Strep. faecium,* and *Strep. durans*) against species of *Clostridium, Bacillus,* and *Lactobacillus* in canned hams was demonstrated by Kafel & Ayres (1969). They believed that enterococci of the types isolated from the hams elaborated an antibacterial metabolite which was active against the susceptible organisms, but activity was also most pronounced at lowered pH values (4.5–6.0). Although the enterococci lowered the oxidation-reduction potential, all the susceptible organisms were ordinarily able to grow at the decreased E_h values. More recently, Jayne-Williams (1973) observed that *Clostridium perfringens* (*Cl. welchii*) was inhibited by *Strep. faecalis* var. *zymogenes*; inhibition was believed to be caused by a bacteriocin.

Earlier work on antagonisms among enteric pathogens and coliforms producing 'colicins' was described by Levine & Tanimoto (1954). Use of antagonistic *Escherichia coli* to inhibit salmonellae in egg white was suggested by Flippin & Mickelson (1960), and Mickelson & Flippin (1960).

Various investigations in our laboratory have been concerned with bacterial associations similar to those described in different types of food products of animal origin. These relations may be of significance with regard to public health and spoilage of the products.

2. Materials and Methods

Several strains of *Strep. thermophilus* and *Lactobacillus bulgaricus* used in yoghurt manufacture were combined in various proportions in sterile milk. Control tubes were inoculated with 1.0% by volume of the inoculum of each organism alone. Differences in coagulation time of the milk were noted, with the results interpreted on the basis of more rapid coagulation occurring as a result of greater acid production. Mixed cultures that caused accelerated acid production in comparison with single culture controls were considered to be stimulatory; if acid production was delayed, the mixture was inhibitory. Various incubation temperatures ranging from 21–48° were tested.

In line with earlier work of Kafel & Ayres (1969), on interactions of bacteria in canned hams, a study was made of inhibitory activity of *Strep. faecalis* isolated from turkey meat against *Cl. perfringens* (ATCC 3624) and *Lactobacillus plantarum* (ATCC 14431) (Borromeo, 1969). Filtrates from the growth medium of *Strep. faecalis* were adjusted to pH 6.7 (the original pH value of APT broth (Difco) used for culture) or left unadjusted. Pure cultures of the test organisms were inoculated into the 2 types of filtrate and into controls of freshly prepared APT broth. Growth was determined by turbidimetric measurements and plate counts after various time intervals. Incubation temperature was 30° for the lactobacillus and 37° for the other organisms.

Later work tested the effects of *Pseudomonas fluorescens* strains isolated from poultry meat on inhibition of 12 *Salmonella* serotypes, *Staphylococcus aureus,* and 2 strains of enterococci (Oblinger & Kraft, 1970). All sensitive organisms were associated with poultry except for *Staph. aureus* 196 E (from M. S. Bergdoll, University of Wisconsin, Madison), and *Staph. anatum* (from the Central Public Health Laboratory, London). In this examination, a perpendicular streak method was used for determining zones of inhibition on brain heart infusion agar (Difco). *Pseudomonas* 'effector' strains were streaked and plates incubated for 48 h at 30° before the sensitive bacteria were streaked perpendicular to the original streak. Several incubation temperatures were tested; 30° for another 24 h was used before measurements were made in cm with a vernier caliper.

Much attention was given in the course of these investigations to the relationship between oxidation-reduction potential and associative growth of several poultry and meat organisms. The system used for determining E_h, pH, and viable cell numbers for *Cl. perfringens* and *Ps. fluorescens* was described by Tabatabai & Walker (1970). This system was modified somewhat in later work by Oblinger & Kraft (1973) on associative growth of *Salmonella* and *Ps. fluorescens.* In this investigation 6 serotypes of *Salmonella* and 3 strains of *Ps. fluorescens* were used. Essentially, measurements were made with appropriate electrodes in an equilibrated system under standardized conditions. Details of the arrangement for making determinations are more completely described in the references cited above.

3. Results

Lactobacilli gain ascendency when grown in mixed culture with an equal inoculum of streptococci in yoghurt manufacture. Therefore, effects of various ratios of rod to coccus were determined for stimulation of acid production. In this work, maximum stimulation occurred with a *Lactobacillus* to *Streptococcus* ratio of 2 cells : 1.

Stimulation was not observed at temperature extremes of 21° or 48°, but greatest stimulation occurred at 37° or 42°. When the total inoculum was varied from 5.6–0.007% with a constant ratio of rod to coccus of 2 : 1, the smallest inoculum produced the longest stimulation, but, as may be expected, a longer time to coagulate the milk than observed with larger inocula.

Filtrates of *Strep. faecalis* grown in APT broth for 6 days had a final pH value of 4.3. These filtrates were inhibitory to *Cl. perfringens* and *Strep. faecalis,* but did not suppress *L. plantarum* completely after 48 h. Adjustment of the pH value to 6.7 allowed growth of all organisms; *Cl. perfringens* was favoured more in steamed than in unsteamed, medium. Presumably, lowering of E_h by steaming was beneficial for the anaerobe. Plate counts of *L. plantarum* and *Strep. faecalis* from the freshly inoculated control APT broth or filtrate, adjusted to pH 6.7, were similar. Cell numbers reached *c.* 1×10^9/ml in both media. However, in the filtrate at pH 4.3, growth was barely supported for the lactobacillus up to 40 h of incubation ($10^2 - 10^3$ cells/ml), while cell numbers decreased for *Strep. faecalis* (from 10^2/ml to < 1/ml). The antagonistic action of the streptococcus reported by Kafel & Ayres (1969) and others appeared to be manifested by lowering of counts of the sensitive *Clostridium* and *Lactobacillus* spp.

Inhibition produced by the 4 strains of *Ps. fluorescens* against several salmonella and other poultry meat isolates is shown in Table 1. Diffusion of

Table 1
Inhibition produced by Ps. fluorescens *strains*

Sensitive organisms	Inhibition produced by *Ps. fluorescens* strains			
	F-21	23a	2	7a
Salmonella strains (12)		2+++		
	6++	5++	4++	
	4+	4+	4+	8+
	2−	1−	4−	4−
Staphylococcus aureus 196E	++	++	++	+
20a	+++	+++	+++	+
Streptococcus faecalis M-18	++	+	++	++
Strep. faecium 7	++	++	++	++
E. coli b3000	++	++	++	+

Zone of inhibition in cm: +, 0.1−1.0; ++, 1.1−1.5; +++, 1.6−2.0. No. of strains tested is prefixed to symbol.

pigment produced by the pseudomonas was associated with a measurable degree of inhibition. Differences in inhibitory activity were readily apparent, but all fluorescent pigment-producers repressed growth of the susceptible organisms to some extent. For all pseudomonas strains, incubation for 36–48 h at 30° provided maximum diffusion of pigment in the agar and resultant inhibition of sensitive bacteria. Several non-pigmented strains were also screened for inhibitory activity. None of these were effective inhibitors. Also, development of inhibition required that the inhibiting pseudomonas strain be allowed a growth advantage of 18–24 h before cross-streaking with the sensitive bacteria. Simultaneous inoculation of both inhibitory and sensitive organisms failed to produce visible inhibition. Competitive growth advantage of pseudomonas strains apparently was not a factor in these tests, but excretion of an antibacterial substance during growth of pseudomonas was responsible for repression of the other bacteria. None of the inhibitory strains of pseudomonas isolated from poultry were mutually repressive.

The significance of oxidation-reduction potential in mutual growth or inhibition of bacteria associated with animal food products has not been fully realized, and we hope that studies described here may awaken interest in this difficult area of research. In the work of Borromeo (1969) on *Strep. faecalis, L. plantarum*, and *Cl. perfringens*, the lower the initial oxidation-reduction potential, the longer was the lag phase for growth of the lactobacillus. This did not occur for the other bacteria, but both *Strep. faecalis* and *Cl. perfringens* in mixed culture were somewhat inhibited in comparison with pure cultures of each. *Lactobacillus plantarum* was inhibited markedly when grown with *Strep. faecalis*; the E_h of the medium was lowered by the streptococcus to levels resulting in greater lag period and lower counts than when *L. plantarum* was grown alone.

Other work in our laboratory (Tabatabai & Walker, 1970) indicated that *Cl. perfringens* strains varied in their response to the initial E_h of the medium. Other

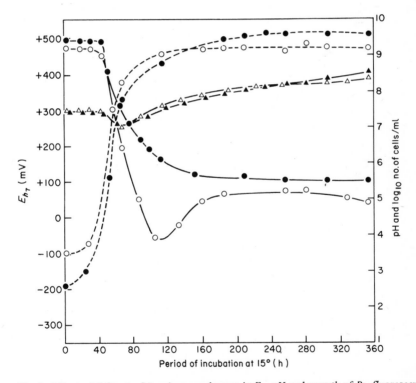

Fig. 1. Effect of *Salm. typhimurium* on changes in E_h, pH and growth of *Ps. fluorescens* F21 in trypticase-soy broth at 15°. Changes are shown in: E_h (● — ●), pH (▲ — ▲) and cell population density (● − − ●) of a pure culture of *Ps. fluorescens* F21. The changes in a mixed culture of the pseudomonad and *Salm. typhimurium*, with an initial ratio of 30 : 1 cells, respectively, are: E_h (○ — ○), pH (△ − △) and population density (○ − − ○).

workers (Kligler & Guggenheim, 1938; Reed & Orr, 1943; Barnes & Ingram, 1956) observed that the limiting E_h for growth of *Cl. perfringens* ranged from −125 to 287 mV (calculated to pH 7.0). One strain of *Cl. perfringens* tested in our laboratory grew better at an initial E_h of +200 mV with small amounts of oxygen present, than at lower E_h of +40 mV in the absence of oxygen. *Pseudomonas fluorescens* at a ratio of 10^6 cells to 10 cells of *Cl. perfringens* in mixed culture had little effect on growth of the anaerobe at initial E_h values of either +200 mV or +40 mV. The pseudomonas could grow in pure culture at an E_h of +40 mV (pH 7.0), but not in the presence of *Cl. perfringens* under similar conditions.

Changes in E_h, pH and cell numbers of *Salmonella typhimurium* and *Ps. fluorescens* F 21 grown together at 15° are shown in Figs 1 and 2. The pseudomonas initially outnumbered the salmonella by a ratio of 30 : 1. Figure 1 shows that the salmonella had little influence on pH, but did cause a marked

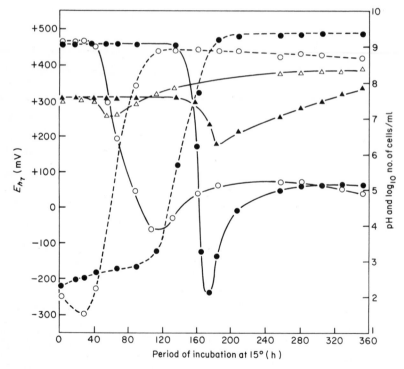

Fig. 2. Effect of *Ps. fluorescens* F32 on changes in E_h, pH and growth of *Salm. typhimurium* in trypticase-soy broth at 15°. Changes are shown in: E_h (● − ●), pH (▲ − ▲) and cell population density (● − − ●) of a pure culture of *Salm. typhimurium*. The changes in a mixed culture of the salmonella and *Ps. fluorescens* F21 with an initial ratio of 30 : 1 cells, respectively, are: E_h (○ − ○), pH (△ − △) and population density (○ − − ○).

decrease in E_h (mixed culture) as compared with the pseudomonads growing in pure culture. Little effect was observed on numbers of cells of *Ps. fluorescens* although the logarithmic and stationary phases of growth were reached somewhat more rapidly in the presence of the salmonella. Figure 2 indicates that the salmonella was influenced by the pseudomonas in mixed culture as shown by more rapid initiation of logarithmic growth (*c.* 80 h sooner than in pure culture), and that the reducing capacity of the salmonella was not as pronounced in mixed culture as in pure culture. The salmonella did not maintain such high cell numbers throughout the incubation period of 360 h in mixed culture as it did when grown alone. However, because growth was started earlier in the presence of the pseudomonas, nutrients may have been made more readily available as a result of the associative growth of the pseudomonads. Utilization of oxygen by the pseudomonads could have encouraged early growth of salmonella. Some evidence of inhibition during the later stages of the stationary period was observed.

4. Discussion

Obviously, the many types of microbial interactions in meats, poultry, and dairy products are too numerous and complex to be discussed adequately here. Compatibility of lactic cultures in yoghurt is one example of an association that is of importance in manufacture of the product. Factors of concern are production of acid and flavour components that are also synchronized with the total manufacturing process. Some flavour results from early development of acetaldehyde by the coccus, *Strep. thermophilus,* with some lactic acid also produced. Then, the typical more highly developed acidity comes about from the activity of the rod, *Lactobacillus bulgaricus.* It then becomes apparent that for successful yoghurt production, both organisms must be present in the proper concentrations at specific times; if one is lacking, the product is defective. The significance of associative growth and activity of these cultures cannot be over-emphasized. This implies proper ratios of inocula in addition to growth conditions favourable to the organisms, and, what is even more important, knowledge of the associative responses of the pair. What causes stimulation or inhibition of one or the other member of the pair is still not well-defined and demands greater study of the nutritional physiology of the bacteria involved.

In considering interactions of organisms found on meats, such as the streptococci, *Cl. perfringens* and lactobacilli discussed here, the changes in the growth medium produced by the dominant organisms should not be overlooked. *Streptococcus faecalis* and other streptococci may produce an antibiotic against the clostridia and lactobacilli which is primarily responsible for their inhibition, but the lowering of pH is a definite factor in antagonism, and changes the composition of the flora of a product such as canned ham. Bacteriocins have been produced by streptococci and their action reported by other workers.

In connection with antibacterial substances produced by *Pseudomonas* spp., pyocins and fluocins as bacteriocins have been shown to cause inhibition of other bacteria. However, it should also be recognized that pseudomonads growing on poultry meat, particularly refrigerated poultry, have a definite advantage over the less numerous pathogenic salmonellae − a very fortunate occurrence from the public health aspect. The fact that salmonellae are still common contaminants of raw poultry does not diminish the concept of the suppression of many potential pathogens by the more competitive psychrotrophic flora.

In all interactions among micro-organisms occurring in foods, some change must be undergone in oxidation-reduction potential of the food substrate. This area should be more thoroughly investigated for further understanding of effects of vacuum- or gas-packaging of meats and poultry on the subsequent microflora of the products. Vacuum packaging is becoming more widespread in meat handling, and information on effects of this practice has not kept pace with its proposed use. Ingram (1962) stated that vacuum packaging of meats lowers the oxidation-reduction potential, but aerobic organisms may still grow. From studies done in our laboratory as well as others, it has been established that there are definite values, or at least ranges, of E_h that determine if a given organism can initiate growth or even survive. In work reported here, a limit of +80 mV (pH 7.0) for *Ps. fluorescens* and +30 mV (pH 7.0) was shown for the *Salmonella* serotypes tested. All the *Salmonella* serotype species had the capacity to bring about intense reducing conditions in a short time. This type of information has some bearing on the effect of packaging conditions on favouring growth of one organism over another, but the importance of these findings needs further examination.

5. Summary

Many investigations have been concerned with microbial interactions from aspects of inhibition or stimulation; several types of associations of micro-organisms are important in foods. Strain selection of starter cultures in the dairy industry is based on compatibility. A survey of strain interactions in yoghurt cultures of *Strep. thermophilus* and *L. bulgaricus* indicated that stimulation of acid production was dependent upon growth response of the cultures to each other, a ratio of rod to coccus of 2 cells : 1, and a favourable temperature.

When *Strep. faecalis* was grown in association with *Cl. perfringens* and *L. plantarum*, the streptococcus inhibited the other bacteria; inhibition may have resulted from lowering of pH. Strains of *Pseudomonas* isolated from poultry meat were inhibitory to *Salmonella*, but needed a growth advantage to cause repression. Investigations on the role of oxidation-reduction potential and mutual growth of these organisms indicated that utilization of oxygen by the pseudomonads was beneficial for growth of *Salmonella. Clostridium perfringens* affected growth of *Ps. fluorescens* in mixed culture when the initial oxidation-reduction potential was low, but not at an initial E_h of 200 mV. With the increase in trend toward vacuum packaging of meats and poultry, the significance of oxidation-reduction potential on associative growth of aerobes and anaerobes warrants further study.

6. Acknowledgements

This paper is Journal Paper No. J-7888 of the Iowa Agriculture and Home Economics Experiment Station, Ames, Iowa. Projects No. 2012, 1749, 1838 and 1896. These investigations were supported in part by PHS research grant FD 00441 from the Food and Drug Administration. The authors thank Joan Andersen and Barbara Hallman for typing the manuscript.

7. References

BARNES, E. M. & INGRAM, M. (1956). The effect of redox potential on the growth of *Clostridium welchii* strains isolated from horse muscle. *J. appl. Bact.* **19**, 117.

BORROMEO, M. C. B. (1969). Effect of fecal streptococci on *Clostridium perfringens* and *Lactobacillus plantarum.* Unpublished M.S. thesis. Ames, Iowa, Library, Iowa State University.

FLIPPIN, R. S. & MICKELSON, M. N. (1960). Use of *Salmonella* antagonists in fermenting egg white. I. Microbial antagonists of *Salmonellae. Appl Microbiol.* **8**, 366.

INGRAM, M. (1962). Microbiological principles in prepackaging meats. *J. appl. Bact.* **26**, 259.

JAYNE-WILLIAMS, D. J. (1973). A medium for overcoming the *in vitro* inhibition of *Clostridium perfringens* by *Streptococcus faecalis* var. *zymogenes* and a note on the *in vivo* interaction of the two organisms. *J. appl. Bact.* **36**, 575.

KAFEL, S. & AYRES, J. C. (1969). The antagonism of enterococci on other bacteria in canned hams. *J. appl. Bact.* **32**, 217.

KLIGLER, I. J. & GUGGENHEIM, K. (1938). The influence of vitamin C on the growth of anaerobes in the presence of air, with special reference to the relative significance of E_h and O_2 in the growth of anaerobes. *J. Bact.* **35**, 141.

LEVINE, M. & TANIMOTO, R. H. (1954). Antagonisms among enteric pathogens *J. Bact.* **67**, 537.

MEERS, J. L. (1973). Growth of bacteria in mixed cultures. CRC Critical Reviews in Microbiology. pp. 139-184. January.

MICKELSON, M. N. & FLIPPIN, R. S. (1960). Use of *Salmonellae* antagonists in fermenting egg white. II. Microbiological methods for the elimination of *Salmonellae* from egg white. *Appl. Microbiol.* **8**, 371.

OBLINGER, J. L. & KRAFT, A. A. (1970). Inhibitory effects of *Pseudomonas* on selected *Salmonella* and bacteria isolated from poultry. *J. Food Sci.* **35**, 30.

OBLINGER, J. L. & KRAFT, A. A. (1973). Oxidation-reduction potential and growth of *Salmonella* and *Pseudomonas fluorescens. J. Food Sci.* **38**, 1108.

REDDY, M. S., VEDAMUTHU, E. R., WASHAM, C. J. & REINBOLD, G. W. (1971). Associative growth relationships in two strain mixtures of *Streptococcus lactis* and *Streptococcus cremoris. J. Milk and Food Technol.* **34**, 263.

REDDY, M. S., VEDAMUTHU, E. R., WASHAM, C. J. & REINBOLD, G. W. (1972). Associative growth studies in three strain mixtures of lactic streptococci. *Appl. Microbiol.* **24**, 953.

REED, G. B. & ORR, J. H. (1943). Cultivation of anaerobes and oxidation-reduction potentials. *J. Bact.* **45,** 309.

TABATABAI, L. B. & WALKER, H. W. (1970). Oxidation-reduction potential and growth of *Clostridium perfringens* and *Pseudomonas fluorescens. Appl. Microbiol.* **20,** 441.

The Role of Micro-organisms in Poultry Taints

J. L. Peel and Jennifer M. Gee

*Agricultural Research Council Food Research Institute,
Colney Lane, Norwich NR4 7UA, England*

CONTENTS

1. Introduction

TAINTS IN POULTRY are due to the presence of chemical substances with offensive sensory properties. These may be extrinsic substances introduced by contamination; alternatively they may be formed in the bird or its immediate environment by chemical reactions and it is the purpose of this contribution to review the role of micro-organisms in these latter processes. Table 1 summarizes documented cases of taints in chickens and turkeys. The types of taint in which micro-organisms have been firmly implicated are so far limited to the 'off' flavours that develop during the storage of poultry carcasses and to the musty taint due to 2,3,4,6-tetrachloroanisole.*

In considering the role of micro-organisms in poultry taints, 3 main questions arise. (1) Are micro-organisms involved? (2) If so, which micro-organisms are involved? (3) What metabolic activities are responsible for the formation of the chemical substances causing the taint? To answer the third question it is necessary to know the identity of the substance causing the taint, and this knowledge also facilitates attempts to answer the first 2 questions. These considerations will be borne in mind in what follows.

2. 'Off' Flavours of Stored Birds

Stored poultry carcasses undergo changes in chemical composition at rates dependent upon the conditions. In the short term, these changes may result in

---

* A new glossary (*British Standards Institute*, BS 5098, 1975) defines a taint as "a taste or odour foreign to the product" and an 'off' flavour is defined, separately as "an atypical flavour (usually associated with deterioration)". In this contribution we have included 'off' flavours within the scope of the term 'taint'.

Table 1
Taints in poultry

Description of taint	Causative substance		References
Micro-organisms implicated 'Off' flavours	Mixed microbial products formed by: (a) intestinal microflora	Chickens	Barnes & Shrimpton (1957) Pippen (1967)
	(b) surface microflora	Chickens & turkeys Chickens	Ayres et al. (1950) Barnes & Thornley (1966)
Musty	2,3,4,6-tetrachloroanisole formed by fungi in poultry litter	Chickens	Engel et al. (1966) Curtis et al. (1972) Curtis et al. (1974)
Micro-organisms not implicated Medicinal/chlorine	Chlorine	Chickens Chickens & turkeys	Ranken et al. (1965) Wabeck et al. (1968)
Disinfectant	3-chloro-2-hydroxytoluene (6-chloro-o-cresol)	Chickens	Patterson (1972)
Mouldy/musty	γ-hexachlorocyclohexane (Lindane)	Chickens	Hixson & Muma (1947) de Lavaur & Carpentier (1968)
Rancid	Lipid oxidation products	Turkeys Chickens & turkeys	Cipra & Bowers (1970) Fishwick & Zmarlicki (1970) Pippen (1967)
Fishy/oily	Components of fish meal, fish oils or linseed oil in diet, or products derived therefrom	Chickens Turkeys	Mostert et al. (1968) Miller & Robisch (1969) Rojas et al. (1969) Atkinson et al. (1972) Pippen (1967) Webb et al. (1973)

what some consumers regard as an improvement in flavour, but this is transient and all birds eventually become unacceptable due to the development of 'off' flavours. The most important factors influencing the shelf life of poultry carcasses are the storage temperature, and whether or not the birds have been eviscerated (Newell, Gwin & Jull, 1948; Elliott & Michener, 1965; Walters 1973). The shelf-life of uneviscerated carcasses increases markedly with decreasing temperature, e.g. Barnes & Shrimpton (1957) reported shelf lives of 2 days at 15° and 30 days at 1° for chickens. Special terminal diets and starvation prior to slaughter prolong shelf life somewhat (Shrimpton, Barnes & Miller, 1958; Shrimpton, 1960). The shelf life of eviscerated carcasses, though improved by cooling, is shorter. At 1° they remain acceptable for 9 to 18 days, depending on a variety of factors, notably the mode of processing and the type of wrapping (Spencer & Stadelman 1955; Essary, Moore & Kramer, 1958). Although evisceration removes the largest reservoir of micro-organisms associated with the live bird, it damages the carcass and thus promotes spoilage due to the surface microflora.

Deterioration of uneviscerated birds at temperatures above 5° is accompanied by 'greening' of the surface areas, first around the vent and on the neck, and later on the back and the ribs. These green areas usually become obvious before 'off' flavours have developed sufficiently for the bird to be rejected and in practice greening is used as an indicator that the birds are unfit for consumption.

The term 'off' flavours therefore embraces a number of overlapping situations, and there are at least 2 distinct processes involved. There is the type of spoilage that occurs in the uneviscerated bird at temperatures > 10° accompanied by greening, and the type that occurs in the eviscerated bird at temperatures between 0 and 5°. In the following sections attention will be focussed on these 2 kinds of spoilage.

3. Tainting of the Uneviscerated Carcass

The 'off' flavours that develop in the uneviscerated bird between 10 and 15° become apparent as an odour on eviscerating the bird prior to cooking, and carry through into the cooked bird both as an abnormal odour and taste. Descriptions such as 'gamey', 'gutty', 'visceral' and 'putrid' have been applied to this taint but the sensory aspects have not been examined systematically and the chemical identities of the volatiles that cause the taint have not been established. No analyses have been reported on carcasses stored to the point of rejection but some information has been obtained by Grey & Shrimpton (1967) on the short term changes in birds hung at 10–15° for 4 days. The volatiles from fresh raw muscle were examined by gas-liquid chromatography and 14 components identified; these were mostly alcohols, aldehydes and ketones but included small amounts of ethane thiol (ethyl mercaptan, $CH_3 . CH_2 SH$) and dimethyl sulphide

($CH_3.S.CH_3$). Several of these compounds were also present in the caecal gases. After death, their concentrations increased in both muscle and caecum and hydrogen sulphide appeared. Grey & Shrimpton (1967) therefore suggested that some of the muscle volatiles, notably ethanol and the sulphur compounds, originate in the caecum and diffuse into the surrounding flesh. Although these experiments were not carried to the point where spoilage was obvious, it is possible that when these processes occur over longer periods they cause tainting of the muscle.

The greening of poultry has been investigated by Pennington & Sherwood (1922) and more recently by Barnes & Shrimpton (1957); the evidence obtained suggests that the green pigments formed are probably sulphaemaglobin and sulphur derivatives of other haem pigments. It is known that such derivatives are formed by the reaction of haem pigments with hydrogen sulphide (Michel, 1938). Barnes & Shrimpton (1957) showed that if the chickens were eviscerated at the first signs of greening the greening process was arrested. When the isolated viscera were held in a closed container the blood vessels turned green and hydrogen sulphide diffusing from the viscera was detected. The muscle of the uneviscerated birds remains virtually sterile during storage, but the microflora of the gut increases, and Barnes & Shrimpton (1957) concluded that greening is due to hydrogen sulphide, formed by intestinal micro-organisms, diffusing into the muscle.

These authors also investigated the intestinal microfloras of the uneviscerated bird. After death, there are 4 main groups of bacteria present, in addition to the non-sporing obligate anaerobes: faecal streptococci, lactobacilli, clostridia, and members of the Enterobacteriaceae, often including *Proteus* spp. The total bacterial count is $10^{10}-10^{11}$/g wet wt in the caecum of the freshly killed bird, and up to 10^6 in the duodenum. During storage at $15°$ the faecal streptococci, the clostridia and the Enterobacteriaceae in the caecum increase whereas the obligate anaerobes and lactobacilli do not. In addition, the population appears to spread up the digestive tract and the count in the duodenum increases markedly. Of these organisms the clostridia and the Enterobacteriaceae, including *Proteus* spp., are the most active producers of hydrogen sulphide under laboratory conditions. The Enterobacteriaceae appear to be more important in the stored bird since most of the clostridia do not grow readily at $15°$. None of these organisms grow well at low temperatures, and this accounts for the fact that spoilage of uneviscerated carcasses below $5°$ is not accompanied by greening. Little is known of the ability of these various bacteria to produce the other volatiles found by Grey & Shrimpton (1967).

The situation with the uneviscerated bird is, therefore, that the sensory and chemical descriptions of the taint are incomplete and that strong circumstantial evidence indicates that spoilage results from the biochemical activities of the intestinal bacteria. There is little information on the role of individual species.

4. Tainting of the Eviscerated Carcass

The 'off' odour of eviscerated birds is different from that associated with greening, and has been described as 'tainted', 'acid', 'sour', 'dish-raggy' and 'sweetly rancid'; at a later stage a pungent ammoniacal odour also appears (Ayres, Ogilvy & Stewart, 1950).

The microflora of eviscerated carcasses and of cut-up portions has been investigated in some detail by Barnes & Thornley (1966) and by Ayres *et al.* (1950), respectively. In both situations 'off' odours became apparent when the surface counts reached *c.* 10^8 bacteria/cm^2. Ayres *et al.* (1950) also stated that by this time, minute bacterial colonies were apparent on the cut surfaces and skin of the birds. Initial surface counts were 10^4–10^5 bacteria/cm^2 in both these investigations whereas carcasses from a modern efficient processing plant rarely give counts $> 10^3$ bacteria/cm^2 (Barnes, pers. comm.). Both groups of workers studied the composition of the bacterial flora with broadly similar results. Barnes & Thornley (1966), using carcasses stored at $1°$, showed that the initial surface flora consists of pigmented and non-pigmented *Pseudomonas* spp., *Alteromonas putrefaciens* (then known as *Pseudomonas putrefaciens* and regarded as an atypical *Pseudomonas* sp.) and oxidase positive and negative *Acinetobacter* spp. The kind of organisms developing during storage is influenced by the manner in which the carcass is wrapped. If wrapped with a film permeable to gases, *Pseudomonas* spp. predominate at spoilage. If, however, the wrapping is impermeable to gases, carbon dioxide accumulates, suppressing *Pseudomonas* spp. and *A. putrefaciens* predominates. Bacterial growth on eviscerated carcasses is mainly on cut muscle surfaces and other damaged areas, such as the feather follicles. This can be demonstrated by spraying the carcass with tetrazolium, which is reduced to give a red pigment wherever the micro-organisms are concentrated.

Barnes & Impey (unpublished data) have investigated the ability of these 4 main groups of psychrophilic spoilage organisms to produce 'off' odours when incubated with minced chicken muscle. The odours produced were described as 'putrid', 'faecal', 'stale', 'silage-like', 'cheesey', and 'urine-like', and suggest that any of these organisms could contribute to the spoilage odour. Other investigations of the effect of psychrophilic bacteria on chicken muscle have been concerned with changes in taste or odour before the onset of spoilage (Griffiths & Lea, 1969; Mast & Stephens, 1972).

Barnes & Melton (1971) have studied the extracellular enzymic activities of several of the psychrophilic spoilage organisms. Many of these bacteria utilize protein, DNA and RNA. The majority attack lipids, the breakdown of which may result in the formation of volatile products that contribute to the 'off' odours. There is evidence that some of these lipid-degrading enzymes are active even at temperatures below freezing; Alford & Pierce (1961) have shown that *Ps.*

fragi has lipolytic activity at $-29°$. It is, therefore, possible that some instances of rancidity recorded in frozen poultry carcasses may be due to the action of enzymes in the frozen state, following heavy microbial contamination prior to freezing.

To summarize the situation with eviscerated carcasses; the sensory and chemical aspects of the taint have received little attention and spoilage appears to be associated with the surface bacteria, several species of which are capable of producing 'off' odours when inoculated into minced chicken muscle. The precise role of the dominant bacteria is not known but several species may be involved.

5. Musty Taint

A 'musty' taint in broiler chickens has been the subject of a detailed investigation by Curtis *et al.* (1972, 1974), following earlier work by Engel, de Groot & Weurmann (1966). The odour has been variously described as 'musty', 'fusty', 'earthy' and 'like old sacks', and is most noticeable to the consumer during or immediately after cooking but in severe cases it may be detected in the uncooked chicken. Engel *et al.* (1966) found such a taint in eggs and poultry meat, and showed that certain batches of litter shavings contained an active principle, which gave rise to the taint; this was later identified as 2,3,4,6-tetrachloroanisole. They also showed that the addition of this compound to the diet of birds resulted in both musty birds and eggs.

Curtis *et al.* (1972, 1974) showed that musty birds contain 2,3,4,6-tetrachloroanisole. This compound was found to have one of the lowest known odour thresholds and concentrations of 5 μg/g of poultry litter are sufficient to taint chickens. Pentachloroanisole was also found in smaller amounts in some birds and has a musty odour, but this compound has a much higher odour threshold and, at the levels encountered, is probably not significant as a cause of taint. Circumstantial evidence suggested that tetrachloroanisole might be formed by the microbiological methylation of 2,3,4,6-tetrachlorophenol, which is an important constituent of wood preservatives (Fig. 1). The latter are applied to

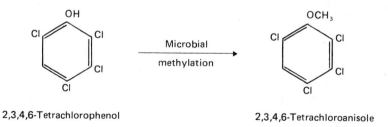

2,3,4,6-Tetrachlorophenol   2,3,4,6-Tetrachloroanisole

Fig. 1. The microbial methylation of 2,3,4,6-tetrachlorophenol.

the surfaces of sawn timber at source to prevent sap-stain and, during subsequent machining of the timber, are largely removed in the shavings. Such shavings form the initial litter in broiler houses and the concentrations of 2,3,4,6-tetrachlorophenol and of pentachlorophenol vary widely from batch to batch (Parr *et al.*, 1974).

The origin of tetrachloroanisole was investigated in a simulated litter system consisting of a sterilized mixture of tetrachlorophenol, water and sawdust, inoculated with a small amount of litter taken from a broiler house associated with musty birds. Subsequent analyses showed that within 10 days most of the tetrachlorophenol had disappeared from the sawdust mixture and had been replaced by an equivalent amount of tetrachloroanisole. Appropriate controls showed that the conversion of the tetrachlorophenol to the tetrachloroanisole could be stopped by heat and was most probably microbiological. Further experiments indicated that the methylating activity could be transferred from one simulated litter mixture to another by small inocula, although it was necessary to supplement the sawdust with a small amount of autoclaved broiler house litter to retain full activity. It would appear that nutrients essential for the growth of the appropriate micro-organisms are present in broiler house litter, probably in the faecal material.

A survey of the microbial flora of broiler house litter was undertaken by Dennis & Gee (1973). The broiler house is a relatively dry habitat, and because certain moulds were already known to carry out analogous methylations, attention was concentrated on the fungi present. Previous studies on the microflora of broiler house litter (Halbrook, Winter & Sutton, 1951; Schefferle, 1965; Lovett, Messer & Read, 1971) were either mainly concerned with bacteria or did not identify fungi beyond generic level. Dennis & Gee (1973) found that the fungal population at the end of an 8–9 week rearing cycle was entirely different from that at the beginning, both in terms of total population and the species represented. *Paecilomyces varioti, Penicillium* spp., *Trichoderma* spp., *Aureobasidium pullulans* and *Hyalodendron lignicola,* were prominent on the fresh shavings but only *Paec. varioti* and the penicilla could be detected in the final litter where *Scopulariopsis brevicaulis* and *Aspergillus* spp. predominated.

The ability of many of these organisms to metabolize and methylate 2,3,4,6-tetrachlorophenol has since been examined (Gee & Peel, 1974). In the test system employed the fungus was first grown for 3–8 days before incubation with tetrachlorophenol. Considerable variations from isolate to isolate were recorded for both of these activities, the 2 extremes being *Penicillium brevicompactum,* where the different isolates utilized between 0 and 100% of the tetrachlorophenol, but where only trace amounts of the tetrachloroanisole were observed, and *Pen. corylophilum,* where high utilization accompanied by high methylation was observed with all the isolates. The bacterial population of litters was not examined in detail, but a mixed bacterial fraction from litter, obtained by filtration and differential centrifuging, metabolized the tetra-chlorophenol completely without methylation. Thus, the disappearance of tetrachlorophenol from broiler house litter probably involves more than one metabolic route.

Although the test conditions did not necessarily reflect those in the broiler house, the ability to methylate tetrachlorophenol is widespread amongst litter fungi. This rules out the possibility that the formation of tetrachloroanisole is due to infection of the litter by a single causative organism, and is in accord with the fact that there is no significant difference between the fungal populations of litters giving rise to musty birds, and those giving rise to normal birds.

The critical factors that determine whether or not the birds become tainted must lie elsewhere. Some light has recently been shed on the situation by a study of the progress of events in broiler house litter (Dennis, Land, Mountford & Robinson, 1975). The fungal population reached a peak after 4–5 weeks and this coincided with the maximum concentration of the tetrachloroanisole. The tetrachloroanisole subsequently disappeared until at the end of the 8–9 week rearing cycle the level had fallen below that associated with tainted birds, and the tetrachlorophenol was at a very low level. If this progression of events were delayed, then the concentration of tetrachloroanisole in the litter at the end of the rearing cycle might be sufficiently high to give musty carcasses. Such a delay may occur, for instance, when the concentration of the tetrachlorophenol in the initial litter is unusually high.

Thus, in the case of this particular musty taint, the chemical identity of the tainting substance is known and its formation from 2,3,4,6-tetrachlorophenol by microbial methylation in poultry litter is firmly established. Although many of the fungal species found in poultry litter are capable of carrying out this methylation, the factors that determine whether a broiler house gives rise to musty birds are not yet clear.

6. General Comments

The poultry taints described here are associated with 3 different microbial populations: (1) the intestinal microflora which gives rise to the taints associated with greening of the carcass; (2) the surface spoilage organisms, which are psychrophilic and give rise to the 'off' taints of eviscerated birds; (3) the organisms of broiler house litter which transform a chemical contaminant of the litter into a tainting compound subsequently taken up by the birds.

In the past 20 years or so, there have been radical changes in the methods of poultry production and processing. Changes in poultry diet, in the conditions of husbandry or in the modes of processing or storage, including the introduction of new chemicals, are all likely to influence the microbial populations of the bird and of its immediate environment. Feed additives or agrochemicals may themselves be metabolized with unforeseen results. The possibility, therefore, always exists that the introduction of new practices may generate new taint problems and these may involve micro-organisms. Beyond these general remarks it is difficult to comment on future trends, since taint problems are not planned but arise unexpectedly.

7. Acknowledgements

The authors wish to thank Dr. Ella M. Barnes and Dr. D. G. Land for helpful discussions and for access to unpublished data.

8. References

ALFORD, J. A. & PIERCE, D. A. (1961). Lipolytic activity of micro-organisms at low and intermediate temperatures. III. Activity of microbial lipases at temperatures below 0° *J. Fd Sci.* **26**, 518.

ATKINSON, A., VAN DER MERWE, R. P. & SWART, L. G. (1972). The effect of high levels of different fish meals, of several antioxidants and poultry byproduct meal on the flavour and fatty acid composition of broilers. *Agroanimalia* **4**, 63.

AYRES, J. C., OGILVY, W. S. & STEWART, G. F. (1950). Post mortem changes in stored meats. I. Micro-organisms associated with development of slime on eviscerated cut-up poultry. *Fd Technol. Champaign* **4**, 199.

BARNES, E. M. & MELTON, W. (1971). Extracellular enzymic activity of poultry spoilage bacteria. *J. appl. Bact.* **34**, 599.

BARNES, E. M. & SHRIMPTON, D. H. (1957). Causes of greening of uneviscerated poultry carcasses during storage. *J. appl. Bact.* **20**, 273.

BARNES, E. M. & THORNLEY, M. J. (1966). The spoilage flora of eviscerated chickens stored at different temperatures. *J. Fd Technol.* **1**, 113.

CIPRA, J. S. & BOWERS, J. A. (1970). Precooked turkey. Flavor and certain chemical changes caused by refrigeration and reheating. *Fd Technol. Champaign* **24**, 921.

CURTIS, R. F., LAND, D. G., GRIFFITHS, N. M., GEE, M. G., ROBINSON, D., PEEL, J. L., DENNIS, C. & GEE, J. M. (1972). 2,3,4,6-Tetrachloroanisole: association with musty taint in chickens and microbiological formation. *Nature, Lond.* **235**, 223.

CURTIS, R. F., DENNIS, C., GEE, J. M., GEE, M. G., GRIFFITHS, N. M., LAND, D. G., PEEL, J. L. & ROBINSON, D. (1974). Chloroanisoles as a cause of musty taint in chickens and their microbiological formation from chlorophenols in broiler house litters. *J. Sci. Fd Agric.* **25**, 811.

DENNIS, C. & GEE, J. M. (1973). The microbial flora of broiler-house litter and dust. *J. gen. Microbiol.* **78**, 101.

DENNIS, C., MOUNTFORD, J., LAND, D. G. & ROBINSON, D. (1975). Changes in the microbial flora, chlorophenols and chloroanisoles in broiler house litter during a chicken rearing cycle. *J. Sci. Fd. Agric.* **26**, 861.

De LAVAUR, E. & CARPENTEIR, J. (1968). Persistance de l'HCH sur différents supports. *Phytiat. Phytopharmacie* **17**, 41.

ELLIOTT, R. P. & MICHENER, H. D. (1965). Factors affecting the growth of psychrophilic micro-organisms in foods. U.S.D.A. Technical Bulletin No. 1320 p. 32. Washington D.C.

ENGEL, C., De GROOT, A. P. & WEURMAN, C. (1966). Tetrachloroanisol: a source of musty taste in eggs and broilers. *Science N.Y.* **154**, 270.

ESSARY, E. O., MOORE, W. E. C. & KRAMER, C. Y. (1958). Influence of scald temperatures, chill times, and holding temperatures on the bacterial flora and shelf life of freshly chilled, tray-pack poultry. *Fd Technol. Champaign* **12**, 684.

FISHWICK, M. J. & ZMARLICKI, S. (1970). Freeze-dried turkey muscle. I. Changes in nitrogenous compounds and lipids of dehydrated turkey during storage. *J. Sci. Fd Agric.* **21**, 155.

GEE, J. M. & PEEL, J. L. (1974). Metabolism of 2,3,4,6-tetrachlorophenol by micro-organisms from broiler-house litter. *J. gen. Microbiol.* **85**, 237.

GREY, T. C. & SHRIMPTON, D. H. (1967). Volatile components of raw chicken breast muscle. *Br. Poult. Sci.* **8**, 23.

GRIFFITHS, N. M. & LEA, C. H. (1969). Chemical and organoleptic changes in poultry meat resulting from the growth of psychrophilic spoilage bacteria at 1°C. 5. Effects on palatability. *Br. Poult. Sci.* **10**, 243.

HALBROOK, E. R., WINTER, A. R. & SUTTON, T. S. (1951). The microflora of poultry house litter and droppings. *Poult. Sci.* **30**, 381.

HIXSON, E. & MUMA, M. H. (1947). Effect of benzene hexachloride on the flavor of poultry meat. *Science N.Y.* **106**, 422.

LOVETT, J., MESSER, J. W. & READ, R. B. (1971). The microflora of Southern Ohio poultry litter. *Poult. Sci.* **50**, 746.

MAST, M. G. & STEPHENS, J. F. (1972). Effects of selected psychrophilic bacteria on the flavor of chicken breast meat. *Poult. Sci.* **51**, 1256.

MICHEL, H. O. (1938). A study of sulfhemoglobin. *J. biol. Chem.* **126**, 323.

MILLER, D. & ROBISCH, P. (1969). Comparative effect of herring, menhaden and safflower oils on broiler tissues fatty acid composition and flavor. *Poult. Sci.* **48**, 2146.

MOSTERT, G. C., DREOSTI, G. M., SWART, L. G., ATKINSON, A. & VAN ZYL, E. (1968). Flavour and growth studies with high levels of South African fish meals in high density broiler rations. *S. Afr. J. agric. Sci.* **11**, 295.

NEWELL, G. W., GWIN, J. M. & JULL, M. A. (1948). The effect of certain holding conditions on the quality of dressed poultry. *Poult. Sci.* **27**, 251.

PARR, L. J., GEE, M. G., LAND, D. G., ROBINSON, D. & CURTIS, R. F. (1974). Chlorophenols from wood preservatives in broiler house litter. *J. Sci. Fd Agric.* **25**, 835.

PATTERSON, R. L. S. (1972). Disinfectant taint in poultry. *Chemy. Ind.* p. 609.

PENNINGTON, M. E. & SHERWOOD, C. M. (1922). The greening of poultry. *Poult. Sci.* **1**, 114.

PIPPEN, E. L. (1967). Poultry flavor. In *Symposium on Foods: The Chemistry and Physiology of Flavors.* Eds H. W. Schultz, E. A. Day & L. M. Libbey. Westport, Connecticut: The Avi Publishing Co. Inc.

RANKEN, M. D., CLEWLOW, G., SHRIMPTON, D. H. & STEVENS, B. J. H. (1965). Chlorination in poultry processing. *Br. Poult. Sci.* **6**, 331.

ROJAS, S. W., LUNG, A. B. & NINO DE GUZMAN, R. V. (1969). Effects of Peruvian anchovy (*Engraulis ringens*) meal supplemented with santoquin on growth and fishy flavor of broilers. *Poult. Sci.* **48**, 2045.

SCHEFFERLE, H. E. (1965). The microbiology of built up poultry litter. *J. appl. Bact.* **28**, 403.

SHRIMPTON, D. H. (1960). Control of greening in undrawn hens. *Agriculture* **67**, 20.

SHRIMPTON, D. H., BARNES, E. M. & MILLER, W. S. (1958). The control of spoilage of uneviscerated poultry carcasses by treatment with antibiotics before slaughter. *J. Sci. Fd Agric.* **9**, 353.

SPENCER, J. V. & STADELMAN, W. J. (1955). Effect of certain holding conditions on shelf-life of fresh poultry meat. *Fd Technol. Champaign* **9**, 358.

WABECK, C. J., SCHWALL, D. V., EVANCHO, G. M., HECK, J. G. & ROGERS, A. B. (1968). Salmonella and total count reduction in poultry treated with sodium hypochlorite solutions. *Poult. Sci.* **47**, 1090.

WALTERS, J. (1973). Food experts tackle EEC health hurdles. *Poult. Wld* **124**, 14.

WEBB, J. E., BRUNSON, C. C. & YATES, J. D. (1973). Effects of feeding fish meal and tocopherol on the flavor of precooked, frozen turkey meat. *Poult. Sci.* **52**, 1029.

Microbiological Hazards of International Trade

BETTY C. HOBBS

*Food Hygiene Laboratory, Central Public Health Laboratory,
Colindale Avenue, London NW9 5HT, England*

CONTENTS

1. Introduction

THE PASSAGE OF FOOD between nations means the sharing of micro-organisms, some of which may be significant agents of disease. The methods of storage and usage of foods containing pathogenic organisms will determine the safety of the food. Sampling, testing, legislative standards and Codes of Practice may help to eliminate the worst supplies, but some contaminated food will reach manufacturers, retail shops and kitchens of canteens and private houses.

Most imported foods in countries such as the United Kingdom, come by sea and air cargo transport. In addition, and the significance cannot be ignored, ferries, cruising ships, railroads and aeroplanes circling the world, pick up foods and set down passengers on a vast international scale. Small amounts of such foods will be taken ashore but much will be ingested and unknown numbers and types of micro-organisms will pass from the food sources to the sewers and general environment of countries other than those preparing the food. Whether these organisms can encroach on the natural home flora will depend on many factors, but mainly on the efficiency of sewage and water purification systems, cleanliness of the general environment and eating habits of the people.

161

The survival of disease producing organisms depends on congenial and susceptible hosts and this in turn may depend on the condition of immune systems in both man and animals. The agents of the zoonotic diseases such as tuberculosis, anthrax, salmonellosis, leptospirosis, brucellosis, psittacosis and others move easily between man and animals so that the importance of animal feedstuffs both imported and home-produced, and of live animals transported for food purposes, cannot be ignored.

The disposal of human and animal waste, both food and faecal, is also a problem and the associated health hazards may be regarded as international as well as national.

2. Seaport Foods

A variety of packed foods from various lands enter other countries through ports. Canned and frozen foods predominate but there are chilled and dehydrated foods and feeds for animals also. There is some information available about these products but it is meagre considering the vast importation of supplies.

(a) Meat

Up to 1967 the frequency of isolation of salmonellae in frozen boneless meat reaching London was known to vary between 'not found' in some batch samples to 100% in others. Since 1968 few samples have been submitted for examination (Hobbs & Wilson, 1959; Hobbs, 1965).

In Glasgow, however, samples are examined from c. 2.5% of cartons in consignments of meat, for example, 30 of 1200 cartons; the 1971–1973 results are given in Table 1. When 1–3 samples examined are positive the particular

Table 1

Salmonella from imported boneless meat (Scotland)

Country of origin	Meat	1971		1972		1973	
		No. examined	No. positive	No. examined	No. positive	No. examined	No. positive
Australia	Beef	145	11(8)	562	4(0.7)	939	21(2)
	Mutton/lamb	–	–	151	4(3)	195	25(13)
New Zealand	Beef	22	0	113	0	153	0
	Mutton	–	–	–	–	55	0
France	Beef	1380	5(0.04)	440	0	59	0
S. Africa	Beef	–	–	22	1(5)	115	0
S. America	Beef	–	–	60	0	30	0
Scandinavia	(?)	70	4(6)	–	–	–	–

(Percentage in brackets)

packs are destroyed and the rest of the consignment is held pending re-sampling. If re-sampling (and, usually, a second re-sampling) proves negative, the consignment is then released for sale, in which case the importer suffers only a negligible loss. If 4–5 or more of the first 30 samples are positive, this is regarded as potentially heavy contamination. Even so, re-sampling is carried out and, if similar results are obtained, the entire cargo is seized. However, if no more positive samples − or, at most, only 1 or 2 out of 60 cartons − are detected on re-sampling and second re-sampling, i.e. 4–7 positives in all out of a total of 90 cartons examined, these 4–7 cartons are destroyed and the remainder of the cargo released for sale.

A cargo, seized as 'unfit for human consumption' under Section 9 of the Food and Drugs Act, is allowed to be sold for the manufacture of pet foods, if the owners or agents agree to the condition that it be sterilized before or during the process of manufacture, and sign an affidavit to that effect. In practice they seldom, if ever, withhold their consent, as this partial loss (to them or their insurers) is more acceptable than the total loss of the value of the cargo which they would otherwise face.

In May 1973 the Medical Officer of Health, and Chief Veterinary Officer, and the Director of the City Laboratory, jointly submitted a recommendation that seized cargoes be released after sterilization for the manufacture of various canned meat products for human consumption. They felt that the present procedure was wasteful of a valuable protein-rich food and that its adequate sterilization would be a safeguard against any public health risk. But their recommendation has not so far been officially sanctioned, perhaps because of the practical difficulties in implementing this policy. Facilities for sterilizing such large consignments of meat are virtually non-existent outside food factories and manufacturers are reluctant to risk contamination of their premises by allowing such cargoes to be delivered even to their sterilizing bays.

It seems that, at present, Glasgow is the only port in the U.K. at which regular sampling of imported raw meat for salmonellae is carried out. Although such random sampling is arbitrary (little more than 2 lb of samples out of a total cargo of 60 000 lb) it provides, at least, a small measure of the general degree of contamination; and magistrates who issue the affidavits accept the results of the examination of these samples as representative of the whole consignment on the authority of their experienced professional advisers.

Inspectors in other ports rely on visual inspection for decomposition, slime and mould. Whereas moulds have been generally regarded as harmless, although they may be indicative of serious bacterial contamination, information is growing about mycotoxins from a variety of fungi in addition to *Aspergillus flavus*; perhaps the hazards of mouldy foods should be taken more seriously.

Horsemeat imported for human consumption is not subjected to heat treatment but horsemeat imported for animal consumption, if not originally produced for human consumption, must be subjected to a process of 'sterilization' *Statutory Instruments* (1969). There are no results available from the examination of this product for salmonellae since 1968 when rates of contamination varied from *c*. 10–100% of samples received. Vast quantities of frozen offal pass through the ports without microbiological examination. Frozen offal which is imported as unfit for, or not intended for, human consumption is subjected to 'sterilization' as unfit meat.

It should be more generally appreciated that a high rate of contamination with salmonellae may be indicative not only of the danger to the community of salmonellosis but also of contamination with other pathogens including viruses, e.g., of foot and mouth disease, when animals or meat come from an area endemic for the disease.

(b) *Poultry*

Poultry carcasses and chicken pieces are habitual sources of salmonellae from both home-produced (Hobbs, 1971; Roberts, 1972) and imported birds. Tables 2–5 give information on the isolation of salmonellae from small numbers of imported samples compared with the results from frozen chicken carcasses and

Table 2
Salmonella from frozen chicken

Country of origin	Year	No. examined	No. positive	No. of serotypes
England	1969–1970	90	22(24.4%)	9
England	1971	100	42(42.0%)	14
England	1972	150	32(24.0%)	9
England	1973	51	25(50.0%)	8
Denmark	1972	132	12(9.1%)	1

Table 3
Salmonella in frozen chicken, 1970–1973

Producer (U.K.)	No. examined	Occurrence of salmonellae in samples		
		No. +ve	% +ve	No. of serotypes
A	136	25	18	9
B	101	53	52	11
C	100	4	4	2
D	56	33	59	2
Others, (9 producers)	102	53	52	1–6

Table 4

Salmonella from frozen chicken pieces

Country of origin	Year	No. examined	No. positive	No. of serotypes
England	1969–1970	100	13(13.0%)	4
Denmark	1969–1970	100	0	
Holland	1973	26 (packets)	17(65.4%) (packets)	3

Table 5

Salmonella from duck carcasses

Country of origin	Year	Incidence of salmonellae in carcasses	
		Dressed	Uneviscerated
United Kingdom	1967–1973	38–78%(55%)	25–71%(37%)
Poland (18/21 exd.)	1973	86%	—

pieces, and frozen ducks produced in this country. The U.K. export trade for frozen poultry ensures further admixture of salmonella serotypes elsewhere. The surveillance results show little improvement in contamination rates over the past 5 years (Table 2) and that only exceptional producers are able to reduce the national level of contamination (Table 3).

The train of events leading to food poisoning in man from poultry include all or some of the following faults: (i) inadequate thawing; (ii) light cooking — enhanced by improper thawing; (iii) cross-contamination to cooked birds from remnants of raw materials on hands, surfaces, receptacles and kitchen tools; (iv) non-refrigerated storage after cooking. Roberts (1972) suggested a temperature of 65° (149° F) for storage of cooked chickens pending hot sale, and Todd & Pivnick (1973) recommended that barbecued food be stored at ⩽ 4.4° or ⩾ 60° (⩽ 40° or ⩾ 140° F).

(c) *Egg products*

Frozen, liquid or dehydrated, bulk egg products (whole egg, or yolk or white) for use in bakeries are frequently contaminated with salmonellae unless the raw product is heat-treated; gastro-enteritis through cross-contamination of foods such as imitation cream with uncooked egg products was common in the past (Hobbs & Smith, 1955; Newell, Hobbs & Wallace, 1955; Smith & Hobbs, 1955). When pasteurized at a suitable time/temperature combination the salmonellae are destroyed (Murdock *et al.*, 1960; Shrimpton *et al.*, 1962; *Anon.* 1963). These products travel in international trade; and those, especially the dehydrated substances which are not known to be pasteurized, should be sampled. For example, some sampling results in 1971 showed that of 102 samples of dried whole egg, 36 (35%), and of 168 samples of dried white, 33 (19%), contained salmonellae. Few results were recorded in 1973; of 66 samples of imported frozen egg albumen, salmonellae were found in 14 (21%).

(d) *Seafoods*

There is a large trade in frozen seafoods, particularly prawns and shrimps, between a number of countries. The microbiological examination usually includes colony and staphylococcal counts as well as a search for salmonellae and vibrios. Table 6 gives results from various countries. An administrative specification in the U.K. currently requires that colony counts at $36 \pm 1°$ be no higher than 10^5-10^6/g and that staphylococcal counts be no greater than 10^3-10^4/g; salmonellae should not be found in 50 g samples. It is expected that the sampling procedure of the International Commission on Microbiological Specifications for Foods (ICMSF) will be adopted (Micro-organisms in Food II: Sampling for Microbiological Analysis — principles and specific applications — in press).

In this, a so-called 'two-class' (pass/fail) plan is recommended for organisms considered hazardous in any circumstances (e.g. salmonellae), and a 'three-class' plan for general colony counts which avoids the 'zero tolerance' of the two-class plan. In the three-class plan, a count m (e.g. 10^4 organisms/g at $36° \pm 1°$) is the level below which it is considered there is no cause for concern, and M (e.g. 10^6 organisms/g at $36° \pm 1°$) is the level beyond which the consignment should be rejected on the basis of risk of disease or spoilage. The lot remains acceptable within a tolerance level c. — the permitted number of sample units (e.g. 2 in 5) having colony counts between m and M. The three-class plan is also used for organisms frequently found in foods but known to cause food poisoning when large numbers are involved (e.g., *Staphylococcus aureus*).

Vibrio parahaemolyticus and non-cholera vibrios as well as salmonellae are also found in frozen cooked prawns from Malaysia (Table 7). It is significant that *V. parahaemolyticus* has been isolated from river water containing effluent from a food factory thawing out large quantities of imported prawns (McCoy, pers. comm.).

(e) *Canned foods*

The classical example of foodborne disease caused by imported canned food is, of course, the outbreak of typhoid fever in Aberdeen (*Anon.* 1964). The can of

Table 6

Frozen cooked peeled prawns (1968–1973)

Country of origin	Years	No. examined	Colony count/g at 35°		E. coli (estimate) $10–10^2$/g	Coagulase-positive staphylococci $> 10^3$/g
			$> 10^5–10^6$	$> 10^6$		
Canada	1968–1973	196	14(7)	0	0	0
Formosa	1969–1973	297	18(6)	0	3(1)	0
Hong Kong	1968–1972	53	33(62)	9(17)	1(2)	0
Japan	1968–1973	608	15(2)	1(0.2)	2(0.3)	1(0.2)
Malaysia	1969–1973	1587	684(43)	171(11)	27(2)	22(1)

(Percentages in brackets).

Table 7
Salmonella and vibrio from Malaysian frozen cooked prawns, 1969–1973

Salmonella sp.	Vibrio parahaemolyticus type*	Non-cholera vibrio*
[19/1687]	[50/872]	[8/872]
Salm. chester (1)	01 : K 26 (2)	Heiberg Group V (4)
Salm. enteriditis (1)	02 : K 28 (1)	
Salm. houten (1)	03 : K 29 (1)	Heiberg Group II (2)
Salm. java phage type Dundee (1)	010 : K 52 (1)	Heiberg Group I (2)
Salm. lexington (1)	Untypable (1)	
Salm. newport (1)	Not typed (44)	
Salm. unnamed (1)		
Salm. weltevreden (12)		

* 1972–1973.

corned beef responsible, undoubtedly heavily contaminated with *Salmonella typhi* phage type 34, came from Argentina and the organism from the sewage polluted water of the River Plate. The water used for cooling the can was assumed to have penetrated through a pinhole leak. Experiments subsequently demonstrated that *Salm. typhi* multiplied readily in sealed cans of corned beef. In mixed inocula the typhoid strain showed strong selective advantage over *Escherichia coli* and *Enterobacter cloacae*. It was also established that the type 34 strain of *Salm. typhi* mutated in corned beef to Vi-type A, which was also isolated from patients in the Aberdeen outbreak (Anderson & Hobbs, 1973). The manner of spread of the organism in the environment of the shop provides excellent teaching material for hygienic studies (*Anon.,* 1964; Couper, Newell & Payne, 1956; Hobbs, 1974). Other similar instances of typhoid fever from canned cured meat are reviewed by Anderson & Hobbs (1973).

Salmonella wien (Wildman, Nicol & Tee, 1951) and other salmonella serotypes have been isolated from canned meat. *Staphylococcus aureus* also has been found in imported canned meat e.g., corned beef. It is possible that this route of importation for intestinal pathogens is more common than appears from the records. Nevertheless, considered in relation to the total imports of canned food, the hazards are small.

Occasionally, fear of *Clostridium botulinum* has called for a search for toxin in certain imports such as canned mushrooms and soups.

(f) *Pharmaceutical products*

There is some traffic in pharmaceutical products, drugs of animal and insect origin, and yeasts. Salmonellae have been found in preparations from the thyroid, pancreas and pituitary glands, edible gelatin, vitamin tablets, lactalbumin and carmine dye. Thyroid extracts, pancreatin, carmine dye and yeast have been reported as sources of human infection in patients undergoing treatment (Williams & Hobbs, 1974).

(g) *Animal feeds*

Large quantities of dehydrated animal and vegetable protein are consumed by birds or animals reared intensively. Although some of this material is manufactured locally from animal waste and grain most of the supplies come from overseas. They include fish meal from Scandinavia, S. America, Canada and S. Africa, bones and bone meal from India, Pakistan, S. America and probably some other countries, and meat meals from N. America; the vegetable meals as such are ground from supplies of grain from the U.S.A., Canada and the U.S.S.R. The grains may have a low salmonella content, sometimes due to dual treatment of animal and vegetable protein in the same factory using the same machinery. On the other hand the animal (meat, bone and offal), fish and poultry (feather, offal and other waste) meals have a variable but often high complement of salmonellae. The degree of contamination depends on the pollution of the raw waste and the hygiene of manufacture, particularly the separation of raw and processed material.

Denmark legislates for the re-sterilization of all imported meat and bone meal (*P.H.L.S. Working Party,* Skovgaard & Nielson, 1972). Marthedal (1973) stated that the Government Circular of 1954 enforcing the heat treatment of meat and bone meal has reduced the introduction of new salmonella serotypes into poultry flocks.

Certain imports into the U.K. have been stopped e.g., sun-dried fish meal from Angola and meat meal from N. America, and pelleting of products has been encouraged for others (Riley, 1969). Pelleting, however, is not the final answer as pellets of fish meal produced by the cold process have yielded a high rate of salmonella isolation (*P.H.L.S. Working Group,* Skovgaard & Nielson, 1972): sufficient heat must be employed. The more enlightened firms have substituted inorganic minerals for bone flour (Riley, 1969).

Nevertheless, the feed meal industry has a difficult task to supply farmers with sufficient quantities of animal feed ingredients. With regard to fishmeal, storms at sea, movement of shoals of fish, fluctuating prices, and shortage of waste materials force the industry to use any supplies that are available. Also the smaller firms buy products rejected, for reasons of contamination, by larger companies. Table 8 is reproduced from a paper (*P.H.L.S. Working Group,* Skovgaard & Nielson, 1972) giving the rate of salmonella contamination of various feeds sampled in the South West of England during the period November, 1968 to January, 1970.

Table 8

Salmonella from different ingredients of pig feed
(England and Wales)

Raw material	No. examined	No. positive	Positive (%)
Feather meal	99	27	27
Raw materials (unspecified)	138	36	26
Meat and bone meal	704	163	23
Fish meal	31	7	23
Fish pellets	264	53	20
Herring meal	60	3	5
Sow nuts	162	1	1

It is now generally recognized that salmonella serotypes introduced in feedstuffs from overseas infect animals, chiefly pigs and poultry, mostly non-clinically. The growth of salmonellae in rehydrated animal feeds has been described (Linton, Jennett & Heard, 1970). Sporadic cases and outbreaks of salmonellosis in man frequently occur because food such as meat and poultry comes from animals excreting or infected with the same serotypes. These foods enter factories, shops and the kitchens of homes and institutions and contaminate the environment, food handler and cooked foods. Rowe (1973) and Lee (1973, 1974) describe the epidemiological events leading from feedstuffs to human infection for *Salm. agona, Salm. 4,12 : d : −, Salm. panama, Salm. virchow* and other serotypes (Table 9).

Salmonellae are not the only microbial hazards in animal feeds and fertilizers. Anthrax bacilli may be present, particularly in crushed bones and bone meal

Table 9

Prevalent salmonella serotypes (up to 1971)

Salmonella sp.	Animal reservoir	Import (suspected)
Salm. virchow	Poultry	Feedstuffs, chicks or eggs
	Porcine	
Salm. 4,12 : d : −	Poultry	Feedstuffs
	Porcine	
	Bovine	
Salm. agona	Poultry	Feedstuffs
	Porcine	
	Bovine	
Salm. panama	Porcine	Bones
	Poultry	Horsemeat
Salm. saint-paul	Poultry	?
	Bovine	

Data from Rowe (1973).

(Enticknap *et al.*, 1968; *Anon.*, 1974a), and anaerobic sporing bacilli are also sometimes present, especially *Cl. welchii*, and even *Cl. botulinum*, both of which are significant pathogens for man and animals. Viruses and toxin-producing fungi may be present also.

(h) *Other foods*

An outbreak of food poisoning due to *Salm. eastbourne* in Canada and the U.S.A. during the autumn and winter period 1973-1974 was associated with chocolate confectionery. One hundred and sixty four cases were reported of which a high proportion were children. Attention is drawn to the salmonella contamination of cocoa bean imports (*Anon.* 1974b).

In the past, dehydrated coconut was a vehicle for salmonellae including *Salm. typhi* and *Salm. paratyphi B* causing food poisoning and enteric fever (Wilson & MacKenzie, 1955). Food-poisoning incidents implicating coconut have not been reported for many years.

Salmonellae were not found in *c.* 166 samples of desiccated coconut submitted by the Port of Liverpool in 1973 and 1974. *Salmonella senftenberg* was isolated from one sample in 1971, this was the last isolation made. Measures taken by the industry in Ceylon have been effective, and perhaps similar revision of procedures are needed for the cocoa bean industry.

3. Animals

Most live animals, such as cattle, pigs, sheep, cats and dogs are subjected to quarantine regulations, although it is doubtful whether the excretion of salmonellae would be observed during the quarantine period. Yet some pets such as terrapins (tortoises or turtles) and exotic birds are allowed into the U.K. without scrutiny. The foods provided for the journey and given on arrival are often meat scraps, a proportion of which are likely to contain salmonellae. Terrapins are well known sources of salmonella and arizona organisms; cases of salmonellosis in young children, and more rarely adults, who handle them occur from time to time (Williams & Helsdon, 1965; Jephcott, Martin & Stalker, 1969; *Anon.*, 1969b; Rowe, 1973). Federal Regulations in the U.S.A. (Wilson, 1972), ban importation of terrapins, turtles and tortoises <6 in. long (*c.* 15 cm). They state that all such reptiles produced domestically and moving in interstate commerce must be certified salmonella-free using approved testing procedures; in certain local areas it is required that a warning sign be placed on turtle tanks in pet shops explaining the possible health hazards (Williams & Hobbs, 1974). Imported exotic birds for which there are no quarantine provisions may be vectors of salmonellae and other zoonotic diseases (e.g. Newcastle disease, psittacosis etc.).

The quarantine arrangements for imports of day-old chicks and hatching eggs for breeding purposes have been revised recently (Animal Health Circular No. 74/67). The period of quarantine has been reduced from 6 to 2 months. Dead birds must be examined in an approved laboratory. These imports have undoubtedly introduced salmonella serotypes into breeding flocks and so to growers for the broiler industry and thus to man. Marthedal (1973) and Lee (1973) describe the paths of spread of infection in the poultry industry. Turkey poults are presumably included under Animal Health Circular No. 74/67.

It may be interesting to speculate whether the importation of antibiotic resistant strains of organisms in man and animals play a significant role in the spread of R factors within a country.

The development of chloramphenicol resistance in *Salm. typhi* (Vázquez *et al.*, 1972; *Anon.*, 1973) is a good example. The transfer of resistance factors from non-pathogenic *E. coli* to *Salm. typhi* has been facilitated by the massive use of chloramphenicol in certain countries because of its cheapness, availability and rapid action. Although this transfer of resistance is important in countries where typhoid fever is endemic, it is unlikely to affect adversely the incidence of the disease in countries such as the U.K. where numbers of cases annually are low, (*c.* 0.3/million of population) and mostly imported. For persons returning from abroad infected with chloramphenicol resistant strains of *Salm. typhi* the prognosis may be serious.

4. Airport Foods

Open pack foods can arrive quickly by air from Europe and they may pass through customs and be distributed to continental-type delicatessen shops

Table 10

Colony counts from cooked meat roll (imported in open pack)

Year	pH value	Colony count/g. (at 35°)		E. coli (estimate) No./g.
		Outside	Inside	
1965	5.8	5.5×10^8	1×10^9	—
	6.4	$> 1 \times 10^9$	2×10^6	—
1966	6.1	5×10^8	< 500	—
	5.8	6×10^8	1×10^6	$> 10^2$ (outside)
1967	6.7	1×10^6	< 500	—
	5.6	3×10^6	1×10^7	$> 10^2$ (outside)

before the results of bacteriological examination are available or even without examination. Results (Table 10) from a series of samples taken from a meat roll coated with dried blood, manufactured in Poland, illustrate the variable range of high and low counts that may be found. The interesting factors relating to this particular meat roll were that (a) improved manufacture markedly reduced the count and (b) even when the counts were $> 10^9/g$ no mechanism was available to reject supplies, on the grounds that the clinical significance of high counts, in the absence of recognized pathogens, was not known.

Large blocks of pâté have sometimes arrived by air unfrozen and unchilled, with high plate counts of mixed organisms including *Staph. aureus*. At first there was no control over transport temperatures and little information about preparation and conditions of storage prior to transport. Complaints and local rejection of a few lots with particularly high counts did not result in marked improvement (Table 11).

Table 11

Colony counts from pâté (imported open pack)

Year	No. examined	Colony count/g at 35°			
		$< 500-10^3$	10^3-10^5	10^5-10^6	$10^6->10^8$
1972	59	28	11	4	16
1973	33	8	5	9	11
Totals	92	36(39)	16(17)	13(14)	27(29)

(Percentages in brackets).
Numbers of potential pathogens were as follows: *Staph. aureus* (3 samples) 7500/g, $2 \times 10^6/g$, 800/g; coliforms (7 samples) $> 10^2/g$; *E. coli* (35 samples) $> 10^2/g$.

If there is a shortage of cold storage space in the air cargo sections of airports the holding of foods in warm weather will aggravate the microbiological hazards.

There appears to be a thriving airport trade in fermented and dried meats and sausages from the continent. The significance of the enormous counts in these products is hard to assess. It must be assumed that fermentation with streptococci, lactobacilli and other organisms has been carried out for years with strains known to be harmless. Table 12 gives some of the results; salmonellae were found in one sample only and the source was investigated.

Shellfish, oysters, mussels and snails may arrive in this country by air; salmonellae were isolated from samples taken from 4 different batches of live snails in sacks from W. Africa and Cyprus, (Table 13).

Table 12

Colony counts from imported continental cured sausage and meats, (1968–1973)

	No. examined	Colony count/g at 35°				
		$< 10^5$	$10^5–10^6$	$10^6–10^7$	$10^7–10^8$	$> 10^8$
Sausage						
Raw*	135	28(21)	20(15)	33(24)	32(24)	22(16)
Cooked	62	24(39)	13(21)	13(21)	12(19)	0
Meat						
Raw	10	4(40)	2(20)	4(40)	0	0
Cooked	52	22(42)	11(21)	9(17)	10(19)	0
Dried	53	1(2)	3(6)	17(32)	19(36)	13(25)

* *Salm. panama* in one sample.
(Percentage in brackets).

Table 13

Salmonella from imported snails, frogs legs and turtle meat
(Food Hygiene Laboratory)

Product	No. examined	No. positive	Serotypes found	Colony count /g at 35°	*E. coli*/g (estimate)
Snails	5	5	typhimurium brancaster enugu deversoir cyprus	250×10^6	$> 10^2$
Frog legs	8	3	thompson java bareilly	5×10^6	$10^1–10^2$
Green turtle	8	8	oslo konstanz arizona derby oranienberg gatuni		

Other open pack foods coming by air and by sea may introduce intestinal pathogens. Frogs legs are usually frozen raw, and salmonellae are found in a proportion of samples. High isolation rates for salmonellae on frogs legs have been reported from the U.S.A., Canada, Greece and France. The exporting countries include France and Indonesia (to the U.K.) and also India, Japan, Pakistan, Mexico and Cuba (*Anon.*, 1969; Williams & Hobbs, 1974). Eight samples from various batches of green-turtle meat from Jamaica all yielded salmonella although in 2 samples the counts were low and coliform bacilli were not found (Table 13).

5. Air Flight Meals

Although traditionally not regarded as import trade, uneaten meals and remains of meals prepared in many countries will be discarded at various airports including London; London alone sends out > 12 million flight meals each year. Waste will be dispersed in a number of ways: to animals, such as pigs; perhaps even to personnel; to sewage lines; and to waste heaps frequented by rats, cats, dogs and insects. Some organisms from the foods that are eaten will join the sewage systems in faeces. The numbers will be small compared with the native flora and probably reduced to infinitesimal numbers before reaching the land or rivers and sea. Those organisms reaching the rivers and sea may be assimilated in shellfish and other seafood creatures.

When food poisoning occurs from in-flight meals due, for example, to *Vibrio parahaemolyticus* (Peffers *et al.*, 1973), the organism will be excreted in large numbers from the stools of the infected individuals and may ultimately reach rivers in the host country. *Vibrio cholerae* El Tor and non-cholera vibrios may likewise travel from country to country; the hazard within a country will depend on the efficiency and safety of sewage and water plants (Sutton, 1974).

Water for flight kitchens and planes, as elsewhere, must be safe for drinking and food preparation. Polluted water used in washing vegetables, dishes, mops or cloths leads to the spread of organisms to the environment and personnel.

The preparation of food in flight kitchens and the loading of chilled meals onto aircraft requires the utmost care in handling and storage; the hazard of bacterial multiplication will increase in hot climates and also by delays in take-off times. Table 14 gives the results of examination of flight meals prepared in London, Europe and the Middle and Far East.

The significance of the high counts depends on safety and spoilage factors which are related to types as well as numbers of organisms.

Table 14
Colony counts from air flight meals, (1971–1973)

Country of origin	No. examined	Colony count/g at 35°					E. coli (estimate) $10–10^2$/g
		<500–10^3	10^3–10^5	10^5–10^6	10^6–10^7	10^7–10^8	
Britain	178	66(37)	51(29)	26(15)	20(11)	15(8)	18(10)
Europe	68	10(15)	25(37)	15(22)	8(12)	10(15)	15(22)
Middle East	47	20(43)	14(30)	3(6)	3(6)	7(15)	14(30)
Far East	26	11(42)	4(15)	3(12)	3(12)	5(19)	4(15)

(Percentage in brackets).

6. Passenger Ships

The potential for spread of micro-organisms between crew and passengers and between food and all personnel on ships cannot be ignored. When certain organisms are dispersed widely amongst passengers by means of food and water from a variety of countries, they may be carried home in large numbers within many people and presumably this could initiate a cycle of infection in the home country. Clinical cases will excrete enormous numbers of infective agents of gastro-enteritis and the gut flora may be slow to return to normal. A continuing incidence of food poisoning on board may enhance the opportunity for a non-indigenous organism to establish itself in the country of disembarkation.

Cholera, the enteric fevers and other salmonella infections will be recognized, and precautions taken to prevent their spread. Other agents such as entero-pathogenic *E. coli*, the lesser known vibrios or even viruses may not be discovered.

In a recent investigation of shipborne gastro-enteritis and galley foods (to be published) high general colony counts, $10^7-10^9/g$, were found in samples of salads, cold meats, cream and mayonnaise, and in some instances there were indications of high *E. coli* counts also. In a series of diarrhoeal cases during a period of *c.* 12 days, *E. coli* of a particular serotype with the same O group (O27) was the predominant organism isolated by direct culture from the faeces of many of the patients. This serotype has not hitherto been associated with food poisoning. It may have been the causative agent in these incidents; the experimental evidence for pathogenicity is now established. Salmonellae and shigellae were not found and although *Cl. welchii* was isolated from a small number of cases there was no common serotype. The incidents occurred sporadically and not as large episodes; influencing factors were probably climatic conditions, foods eaten and the resistance of the diners to infection. It was assumed that the *E. coli* serotype responsible entered the ship in raw foods picked up from various countries, or in members of the Asian galley staff and that it did not constitute part of the normal flora of U.K. persons.

Salmonellosis on ships is not uncommon. Passengers have become infected with serotypes thought to originate from frozen dressed poultry or raw meats taken on board at both home and overseas ports.

Methods of preparation, storage and galley hygiene are all important factors in avoiding spread and build-up of intestinal pathogens passing from raw produce into cooked foods ready for eating.

7. Discussion

The people of privileged countries expect safe food; most have been taught the principles of food hygiene and know the hazards involved. Assuming a nation can exert control over its own bacterial population, what action can be taken when there is dependence on foods from other countries less aware of the need, or less able, to control contamination?

For many years food legislation has been carried out with the help of the Food and Drugs Act, Imported Food Regulations, Food Hygiene Regulations and the various laws associated with particular liquids and foods such as milk, meat and egg mixes. There has been a great measure of success in the application of safety regulations for water, milk, ice-cream and egg products, which have eliminated cholera and markedly reduced the incidence of the enteric fevers, salmonellosis, bovine tuberculosis and brucellosis. Nevertheless, there is still much food poisoning which is relatively unchecked by the present measures of food control. The incorporation of preventive measures against salmonellosis and other diseases from animal sources may be seen in the legislation for milk and egg products, but the known dangers of infective agents spreading from animal feeds, raw meat and poultry products, and from live animals not subjected to quarantine regulations, are not subject to legislative or even administrative action.

The distribution of contaminated raw and cooked products is a constant source of infection from which outbreaks and sporadic cases of food poisoning arise. Regular sampling would select the worst supplies and if criteria for acceptance and rejection could be agreed by all importing authorities, improved protection would be expected. The ICMSF has prepared sampling data and microbiological standards for many foodstuffs common in international trade.

The formulae for sampling and assessment of criteria depend on the numbers of samples/lot (usually 5 or 10), minimum and maximum levels of acceptance and the tolerances allowed. The principle can be applied to the usual microbiological tests, colony counts and particular organisms of significance, for example salmonellae and staphylococci. The specifications for various foods as they will appear in the publication Micro-organisms in Food II: 'Sampling for Microbiological Analysis — principles and specific applications' (to be published) are almost too lenient in many instances, so as not to embarrass the food industry or endanger food resources for the world. Nevertheless, they provide international yardsticks of microbiological cleanliness and safety for many foods dependent on non-sterile processes for production.

There is a fallacy that raw foods, such as meat and poultry which must be cooked before eaten, can be disregarded as sources of infection; they are excluded from the Food and Drugs Act except on macroscopic appearance and smell indicating obvious spoilage. The hazards of cross contamination from raw to cooked food in preparation areas where both products are handled in close proximity are not always recognized; yet the source of many outbreaks of salmonellosis and *Vibrio parahaemolyticus* food poisoning has been traced to raw materials originating from live animals or sea creatures. Likewise the significance of high colony counts in prepared foods ready for consumption is not recognized by health authorities.

'Travellers' diarrhoea' is precipitated by the high bacterial counts which develop in foods standing at ambient temperatures. The critical clinical levels of food poisoning organisms in foods will be reached faster when surrounding temperatures are high. Furthermore, organisms such as *E. coli* may be enteropathogenic in large doses. So far, certain serotypes only of *E. coli* have been found to cause diarrhoea in adults; perhaps this mainly occurs when non-immune individuals acquire a large dose of such serotypes circulating in countries other than their own. Rowe, Taylor & Bettelheim (1970) and Rowe (1973) describe 0148K?H28 carried in food and by flies as the causative agent of diarrhoea in successive batches of men arriving in army camps in Arabia.

Organisms in the *Pseudomonas, Citrobacter,* and *Providencia* groups and aerobic and anaerobic sporing bacilli in addition to *Bacillus cereus* and *Cl. welchii* may also be capable of causing food poisoning by invasion or by toxin production if consumed in large numbers.

The examination of imported food products is desirable on a far larger scale than is practised at present. Facilities for food microbiology, which requires more apparatus and is more time consuming and perhaps less rewarding than clinical microbiology, are deficient in most countries. Largely they are confined to industrial laboratories monitoring factory products or to those required by large chain stores.

The Public Health Laboratory Service (PHLS) laboratories concerned with imports supply facilities for the examination of port samples, and the inland laboratories assist with foods passing through ports in container transport. However, the work is not carried out on a control basis and it is only a fraction of the general clinical work carried out by these laboratories. If more fundamental sampling were to take place, increased facilities and staff would be required for the PHLS, unless a government food control centre such as that provided by the Food & Drug Administrations in the U.S.A. and Canada were set up. Nevertheless, in view of transport and staffing difficulties, it would be better to encourage an increase in the intake of the existing efficient public health laboratories.

There is no doubt that the importation of contaminated food constitutes a hazard to the population and efforts should be made to monitor the microbiological content of both raw and processed materials.

International Committees including Codex Alimentarius, ICMSF, International Standards Organization and the European Economic Community are working on the production of practicable standards for sampling and examination as well as microbiological specifications. As far as salmonella is concerned there is an urgent need for assessment and comparison of the costs of investigating outbreaks with the costs of eliminating the organisms at source, principally by treatment of feeds and the closer monitoring of imported animals and feedstuffs.

8. Acknowledgements

I am indebted to Dr. T. F. Elias-Jones and Mr. Carlyle McCance for the account of their sampling procedure in Glasgow and for Table 1. Also to the many Environmental Health Officers, especially those of the sea and air ports, who have provided samples for surveillance studies on foods. Mr. Bailey and his colleagues have submitted many flight meals from kitchens all over the world. My loyal, hardworking colleagues in the Food Hygiene Laboratory have carried out the work and I thank them. I am grateful to Miss N. Cockman for collating and summarizing much data, to Dr. J. Lee for help with results from flight meals, and to Dr. P. C. B. Turnbull for editorial help.

9. References

ANDERSON, E. S. & HOBBS, B. C. (1973). Studies of the strain of *Salmonella typhi* responsible for the Aberdeen typhoid outbreak. *Israel J. med. Sci.* **9**, 162.

ANON. (1963). Statutory Instruments. No. 1503, The Liquid Egg (Pasteurization) Regulations. London, HMSO.

ANON. (1964). The Aberdeen typhoid outbreak, 1964. Departmental Committee Report. Cmnd. 2542. Edinburgh, HMSO.

ANON. (1969). Tortoises, Terrapins and Turtles. *Br. med. J.* **iv**, 758.

ANON. (1969). Statutory Instruments. No. 871, The Meat (Sterilization) Regulations. London, HMSO.

ANON. (1969). Salmonella contamination in frog legs. Salmonella Surveillance Report, National Communicable Disease Center, Atlanta, Georgia. No 81.

ANON. (1973). Follow-up on chloramphenicol-resistant *Salmonella typhi* – Mexico. *Morbidity & Mortality,* **22**, 159.

ANON. (1974*a*). Food-borne disease: Methods of sampling and examination in surveillance programmes. *Wld. Hlth. Org. techn. Rep. Ser.* No. 543.

ANON. (1974*b*). Outbreak of food poisoning due to *Salmonella eastbourne*. U.S. Morbidity and Mortality Weekly Report. **23**, Nos 4, 5, 9 & 10.

COUPER, W. R. M., NEWELL, K. W. & PAYNE, D. J. H. (1956). An outbreak of typhoid fever associated with canned ox-tongue. *Lancet,* **i**, 1057.

ENTICKNAP, J. B., GALBRAITH, N. S., TOMLINSON, A. J. H. & ELIAS-JONES, T. F. (1968). Pulmonary anthrax caused by contaminated sacks. *Br. J. Ind. Med.* **25**, 72.

HOBBS, B. C. (1965). Contamination of meat samples. *Mon. Bull. publ. Hlth. Lab. Serv.* **24**, 123, 145.

HOBBS, B. C. (1971). Food poisoning from poultry. In *Poultry Disease and World Economy.* Eds R. F. Gordon & B. M. Freeman. Edinburgh: Longman Group Ltd. for British Poultry Science Ltd.

HOBBS, B. C. (1974). In *Food Poisoning and Food Hygiene.* 3rd ed. London: Edward Arnold (Publishers) Ltd.

HOBBS, B. C. & SMITH, M. E. (1955). Outbreaks of paratyphoid B fever associated with imported frozen egg. II. Bacteriology. *J. appl. Bact.* **18**, 471.

HOBBS, B. C. & WILSON, J. G. (1959). Contamination of wholesale meat supplies with salmonellae and heat-resistant *Cl. welchii. Mon. Bull. publ. Hlth. Lab. Serv.* **18**, 198.

JEPHCOTT, A. E., MARTIN, D. R. & STALKER, R. (1969). Salmonella excretion by pet terrapins. *J. Hyg. Camb.* **67**, 505.

LEE, J. A. (1973). Salmonellae in poultry in Great Britain. In *The Microbiological Safety of Food*. Eds B. C. Hobbs & J. H. B. Christian. London: Academic Press.

LEE, J. A. (1974). Recent trends in human salmonellosis in England and Wales: the epidemiology of prevalent serotypes other than *Salmonella typhimurium*. **72**, 185.

LINTON, A. H., JENNETT, N. E. & HEARD, T. W. (1970). Multiplication of *Salmonella* in liquid feed and its influence on the duration of excretion in pigs. *Res. vet. Sci.* **11**, 452.

MARTHEDAL, H. E. (1973). The occurrence of salmonellosis in Poultry in Denmark 1935-1971 and the eradication programme established. In *The Microbiological Safety of Food*. Eds B. C. Hobbs & J. H. B. Christian. London: Academic Press.

MURDOCK, C. R., CROSSLEY, E. L., ROBB, J., SMITH, M. E. & HOBBS, B. C. (1960). The pasteurization of liquid whole egg. *Mon. Bull. publ. Hlth. Lab. Serv.* **19**, 134.

NEWELL, K. W., HOBBS, B. C. & WALLACE, E. J. G. (1955). Paratyphoid fever associated with Chinese frozen whole egg. *Br. med. J.* **ii**, 1296.

P.H.L.S. WORKING GROUP, SKOVGAARD, N. & NIELSON, B. B. (1972). Salmonellas in pigs and animal feeding stuffs in England and Wales and in Denmark. *J. Hyg. Camb.* **70**, 127.

PEFFERS, A. S. R., BAILEY, J., BARROW, G. I. & HOBBS, B. C. (1973). *Vibrio parahaemolyticus* gastroenteritis and international air travel. *Lancet* **i**, 143.

RILEY, P. B. (1969). Salmonella infection: The position of animal food and its manufacturing process. In *Bacterial Food Poisoning*. Ed. J. Taylor. London: The Royal Society of Health.

ROBERTS, D. (1972). Observations on procedures for thawing and spit-roasting frozen dressed chickens, and post-cooking care and storage: with particular reference to food-poisoning bacteria. *J. Hyg. Camb.* **70**, 565.

ROWE, B. (1973). Salmonellosis in England and Wales. In *The Microbiological Safety of Food*. Eds B. C. Hobbs & J. H. B. Christian. London: Academic Press.

ROWE, B., TAYLOR, J. & BETTELHEIM, K. A. (1970). An investigation of travellers' diarrhoea. *Lancet*, **i**, 1.

SHRIMPTON, D. H., MONSEY, J. B., HOBBS, B. C. & SMITH, M. E. (1962). A laboratory determination of the destruction of α-amylase and salmonellae in whole egg by heat pasteurization. *J. Hyg. Camb.* **60**, 153.

SMITH, M. E. & HOBBS, B. C. (1955). Salmonella in Chinese frozen egg. *Mon. Bull. publ. Hlth. Lab. Serv.* **14**, 154.

SUTTON, R. G. A. (1974). An outbreak of cholera in Australia due to food served in flight on an international aircraft. *J. Hyg. Camb.* **72**, 441.

TODD, E. & PIVNICK, H. (1973). Public health problems associated with barbecued food. A review. *J. Milk Technol.* **36**, 1.

VÁZQUEZ, V., CALDERÓN, E. & RODRÍGUEZ, R. S. (1972). Chloramphenicol-resistant strains of *Salmonella typhosa*. *New Engl. J. Med.* **286**, 1220.

WILDMAN, J. H., NICOL, C. G. & TEE, G. H. (1951). An outbreak due to *Salmonella wien*. *Mon. Bull. publ. Hlth. Lab. Serv.* **10**, 190.

WILLIAMS, L. P. Jr. & HELSDON, H. L. (1965). Pet turtles as a cause of human salmonellosis. *J. Am. med. Ass.* **192**, 347.

WILLIAMS, L. P. Jr. & HOBBS, B. C. (1974). Enterobacteriaceae infections. In *Diseases Transmitted from Animals to Man*. 6th ed. Eds W. T. Hubbert, W. F. McCulloch & P. R. Schnurrenberger. Springfield, Illinois: Charles C. Thomas.

WILSON, M. M. & MACKENZIE, E. F. (1955). Typhoid fever and salmonellosis due to the consumption of infected desiccated coconut. *J. appl. Bact.* **18**, 510.

WILSON, V. E. (1972). Turtles, tortoises and terrapins. *Fed. Reg.* **37**, 7005.

Vibrio parahaemolyticus and Seafoods

G. I. BARROW AND D. C. MILLER

*Public Health Laboratory, Royal Cornwall Hospital (City),
Infirmary Hill, Truro TR1 2HZ, Cornwall, England*

CONTENTS

1. Introduction

ALTHOUGH LARGE quantities of seafoods are consumed annually throughout the world, marine bacteria have been regarded more as spoilage organisms rather than as a cause of illness in man. One marine organism, however, was isolated in Japan in 1951 from cases of acute gastroenteritis by Fujino and his colleagues (1953). This organism, *Vibrio parahaemolyticus,* is now recognized there as the commonest cause of food poisoning in summer due mainly to the national custom of eating raw and semi-processed marine products. Although this halophilic, Gram negative, chitin-utilizing marine organism was first described more than 20 years ago, it has remained practically unknown elsewhere until comparatively recently, probably because most of the early and much of the subsequent work was published in Japanese. At first, it was thought to be limited to Japan and the Far East, but during the last 5 years it has been isolated from numerous marine sources in many countries throughout the world. In some of these countries, food poisoning caused by *V. parahaemolyticus* has now been recognized for the first time. As it may also be imported, not only by travellers but also commercially in seafoods, it is important that both medical and food microbiologists should be aware of the potential significance of this organism. Its importance may be judged by the contributions published in the Proceedings of the International Symposium on *Vibrio parahaemolyticus* (1974) held in Tokyo in 1973. Their wide range summarizes well the direction of current work and thought on this organism.

2. Classification

The organism first isolated from a large outbreak of gastroenteritis was named *Pasteurella parahaemolytica* by Fujino *et al.* (1953). A similar organism was later isolated from salt-media during an outbreak of food poisoning by Takikawa (1958) who called it *Pseudomonas enteritis,* and who subsequently confirmed its enteropathogenicity in human volunteers. It was later thought to be an aeromonad by Miyamoto, Nakamura & Takizawa (1961) who suggested a new genus *Oceanomonas* for it. However, after extensive taxonomic studies, it was placed in the genus *Vibrio* by Sakazaki, Iwanami & Fukumi (1963) who proposed the name *Vibrio parahaemolyticus* which is at present accepted.

Vibrios previously included a wide variety of organisms and this created confusion both in the laboratory and in epidemiological work. During recent years, however, considerable progress has been made in classification and the genus *Vibrio* is now restricted to motile, aerobic and facultatively anaerobic, non-sporing, Gram negative rods with a single polar flagellum, which are oxidase and catalase positive; utilize some sugars fermentatively without gas formation and which produce lysine and ornithine decarboxylase but not arginine dihydrolase (Hugh & Sakazaki, 1972). With newer sophisticated methods of bacterial analysis as well as electron microscopy, DNA base ratios, recombination studies, fatty acid profiles and other techniques, the current taxonomic position of this organism is less certain. For example, under some cultural conditions, peritrichous flagella can be demonstrated in *V. parahaemolyticus* (Yabuuchi *et al.,* 1974), which would thus exclude it from the genus *Vibrio* as currently defined. Baumann, Baumann & Mandel (1971) would place it in the genus *Beneckea* since it hydrolyses chitin, as do the other members of this group. Indeed, Chatterjee (1974) would create a new genus for it intermediate between *Pasteurella* and *Yersinia.* However, in order to avoid further confusion, it was agreed during discussion at the International Symposium to follow the forthcoming edition of Bergey's Manual of Determinative Bacteriology (1974) in retaining the name *Vibrio parahaemolyticus* at present, and this term should therefore continue to be used meanwhile.

3. Isolation and Identification

There are few species within the genus *Vibrio* as currently defined which are known to affect animals; of those which affect man, *V. cholerae* is already well known and *V. parahaemolyticus* is now becoming more widely known. With the spread of the El Tor cholera vibrio in Europe (Lorenzo *et al.,* 1974), it is fortunate that *V. parahaemolyticus* can be isolated by the same media and methods as those used for *V. cholerae.* Like other enteric organisms, they grow well on ordinary laboratory media, but the recent introduction of more selective media have rendered isolation of these and other vibrios much easier. These media include thiosulphate-citrate bile-salt sucrose (TCBS), bromothymol blue-salt-teepol (Sakazaki, 1969) and vibrio agar (Tamura, Shimada & Prescott, 1971) for direct plating and for subcultures, as well as salt-colistin broth, glucose-salt-teepol broth (Sakazaki, 1969, 1973*b*) and Monsur's medium (1963) for enrichment culture. Like alkaline peptone water, all these media depend partly on a high pH value for their selective action, and current trends suggest that this may be further enhanced by incubation at 43°, by the addition of chitin, starch, or mineral salts as well as by the use of fish based media. Absence of growth on salt-free media is useful for confirming that organisms are halophilic.

Vibrio agar and bromothymol blue-salt-teepol agar are not as inhibitory as TCBS medium, but after overnight incubation on the latter the majority of Gram positive organisms, enterobacteria and many proteus strains either do not grow or only form tiny colonies. The characteristic large, green, non-sucrose fermenting colonies of *V. parahaemolyticus* are readily distinguishable from the yellow sucrose fermenting colonies of *V. cholerae* and those of the common marine organism *V. alginolyticus*. *Vibrio alginolyticus* was previously called *V.*

Table 1
Characteristics important for the identification of V. parahaemolyticus

Test	Reaction
Oxidase production	+
Catalase production	+
Inhibition of growth by methylene blue	+
Inhibition of growth by vibriostatic agent O/129	+
Motility	+
Sucrose fermentation*	−
Mannitol fermentation	+
Growth in 1% tryptone broth overnight shake cultures at 37°	
(1) without added NaCl	−
(2) with 6% NaCl	+++
(3) with 8% NaCl	++
(4) with 10% NaCl	−
Growth in 1% tryptone broth with 2% NaCl at 42–43° overnight	+++
Acid without gas from glucose anaerobically in Hugh & Leifson's O−F test	+
Urease production*	−
Voges-Proskauer test in semi-solid medium	−
Indole production*	+
H₂S production	−
Lysine decarboxylase production†	+
Ornithine decarboxylase production	+
Arginine dihydrolase production†	−
Gelatin liquefaction	+
Nitrate reduction	+

* Occasional strains give different results.
† Taylor's modification: peptone omitted from Falkow's medium.
N.B. Except where stated, these tests are performed with media containing 2–3% NaCl.

Table 2

Differentiation of V. parahaemolyticus *from other organisms*

Organism	Colonial appearance on TCBS	TSI slant/butt	Lysine decarbo-xylase	Voges-Proskauer reaction	Growth in 8% NaCl
V. parahaemolyticus	Large, dark green centre	Alk/Acid	+	−	+
V. cholerae	Medium yellow	Acid/Acid	+	d or +	−
V. alginolyticus	Large yellow	Acid/Acid	+	+	+
V. anguillarum	No growth	Acid/Acid	−	+	−
Other marine vibrios	Large green or yellow	d/d	d	−	−
Aeromonas spp.	Small yellow or no growth	Acid/Acid (gas d)	−	d	−
Pseudomonas spp.	Small, pale green or colourless	Alk/Alk	−	−	−
Proteus spp.	Small, yellow, black or green	Alk/Acid (d)/(gas d)	−	− (P. mirabilis, d)	−
Plesiomonas spp.	No growth	Alk/Acid	+	−	−

d, variable.

parahaemolyticus biotype 2, and this may perhaps have been the cause of some confusion in correct identification (Sakazaki, 1968). Whatever techniques used, and whether sucrose fermenting or not, any isolates which are motile, oxidase and catalase positive and sensitive to methylene blue and vibriostatic agent O 129 (Bain & Shewan, 1968) should be regarded as vibrios and further identified (Barrow & Miller, 1972). For *V. parahaemolyticus*, triple sugar-iron agar and sulphide-indole motility medium are useful for initial screening of isolates before further differential tests are performed. The minimum practical characteristics essential for the identification of *V. parahaemolyticus* and its differentiation from some other organisms are shown in Tables 1 and 2.

For the isolation of *V. parahaemolyticus* and other vibrios from seafoods or marine sources enrichment techniques, with subculture to solid selective media, are widely used. It should be noted, however, that marine vibrios other than *V. parahaemolyticus* grow better on bromothymol blue-salt-teepol agar than on the more inhibitory TCBS medium. Much of the work so far published has been concerned with isolation rather than enumeration, but as recently emphasized at the International Symposium (1974), more quantitative work on the microbiology of seafoods is needed, particularly as considerable variation may occur in different parts of the world and with different methods of processing. Such studies depend mainly on dilution techniques in enrichment media or buffered salt solutions, with subsequent subculture to selective solid media, including the anaerobic starch fermentation plate of Baross & Liston (1968). Membrane filtration methods, with or without dilution, are suitable for the liquid from thawed seafoods, or from environmental specimens such as sea water. Each method has advantages and disadvantages — which may vary with geography and climatic conditions — and much current work is directed towards improving media and methods for both isolation and enumeration.

4. Ecology

Vibrio parahaemolyticus belongs to a group of halophilic vibrios, both psychrophilic and mesophilic, which are common in shallow coastal and estuarine waters rather than deep seas. The incidence of these organisms in the marine environment varies with temperature: below *c.* $10°$ they can only be isolated in small numbers from water, but they probably survive in the bottom sediment. Their growth in warm waters and survival in cold conditions is probably closely associated with chitin and the life cycle of zooplankton (Kaneko & Colwell, 1973), although certain industrial effluents as well as commerce in marine products, including animal foods and agricultural fertilizers, may influence both the incidence and actual type distribution (Barrow, 1973*a*; Kristensen, 1974). Some marine organisms are closely related to *V. parahaemolyticus*, and it is therefore essential that all suspect isolates, even on selective media, should be adequately identified biochemically. Failure to do this has been the cause of much contradictory evidence in the literature.

In their natural marine environment, halophilic vibrios constitute a heterogeneous group so that comparison of one with another may be difficult. In the human environment, however, such mixed populations rarely occur and when they do, are less likely to be differentiated from the dominant organism during infection. None the less, as knowledge about *V. parahaemolyticus* has increased, so have the number of 'atypical' strains causing infection, thus indole negative strains, urea positive strains and sucrose positive variants have all been described (Zen-Yoji *et al.,* 1973; Sakai, Kudoh & Zen-Yoji, 1974).

Current work in ecology and numerical taxonomy suggests that, in the marine environment, exchange of genetic material amongst marine vibrios and other bacteria, by conjugation or transduction by bacteriophages, could occur and may well be aided by the concentration of marine organisms in filter-feeding shellfish. In this way, it is suggested, *V. parahaemolyticus* and other vibrios could adapt quickly and efficiently to rapid intertidal changes of organic content, zooplankton concentration, salinity and other factors (Liston, 1973). In this process of evolution, it is not surprising that 'typical' strains of *V. parahaemolyticus* should be outnumbered by parahaemolyticus-like organisms — whatever they may be called in the future — with some differences from those currently associated with human infection. The greater the number of characters considered, the greater the number of differences will be detected, but for practical purposes, those shown in Table 1 are essential for identification. It seems predictable, however, that in time, human infection with less typical organisms will occur, provided that the basic mechanism of pathogenicity is retained.

5. Pathogenicity

In man, *V. parahaemolyticus* usually causes either diarrhoea, occasionally dysentery-like, or gastroenteritis of sudden onset, varying from mild to severe. In contrast to the effortless vomiting and painless diarrhoea of classical cholera, *V. parahaemolyticus* infection is usually accompanied by severe abdominal pain. The average incubation period is 12–24 h, but ranges from 2–48 h, depending partly on the infecting dose, the nature of the food and the condition of the stomach. As in cholera, the vibrios multiply rapidly in the gut and are excreted in large numbers during illness, but they decrease rapidly with clinical recovery. It is self-limiting, generally lasting only a few days, with little evidence of spread of the infection from one person to another. Treatment is preferably supportive, with fluid replacement if necessary; if severe, tetracycline, neomycin or streptomycin may be used, but the vibrio is resistant to ampicillin. *Vibrio parahaemolyticus,* or closely related organisms, have also been isolated occasionally from infected skin or tissue lesions in bathers and fish handlers (Roland, 1970), but its possible pathogenicity for fish, shellfish and crustacea is uncertain.

Like *Salmonella* spp. and *Escherichia coli,* strains of *V. parahaemolyticus* may be differentiated currently into some 54 different serological types by agglutination tests with specific O and K antisera as shown in Table 3 (Sakazaki, Iwanami & Tamura, 1968; Sakazaki, 1973*a*). This antigenic scheme will probably soon be extended. The H antigens of the different strains so far examined are regarded as serologically identical. Some antigens are shared with other marine bacteria and serology cannot be used for identification. The present Japanese serological typing scheme is based only on isolates from cases of human infection, and for this reason, the incidence of as yet untypable isolates from marine sources and seafoods may vary considerably from one area to another. This is a limiting factor in epidemiological work.

Certain serotypes may be isolated more frequently than others, although the actual serotypes may vary with time and place (Kudoh *et al.,* 1974). Whatever the serotype, however, most strains isolated from patients are able to produce β-lysis of human blood in Wagatsuma's agar medium, whereas the majority of isolates from marine sources are usually unable to do so, even when the food is incriminated on epidemiological grounds with illness (Sakazaki *et al.,* 1968; Miyamoto *et al.,* 1969). This test is known as the Kanagawa reaction, so-called from the Prefecture in Japan where it was first developed. These findings accord

Table 3

Serological types of Vibrio parahaemolyticus

O group	K types
1	1; 25; 26; 32; *33; 38; 41; 56; TNK 11.**
2	3; 28.
3	*4; 5; 6; 7; 29; *30; 31; *33; 37; 43; 45; 48; *51; 54; 57.
4	*4; 8; 9; 10; 11; 12; 13; 34; 42; 49; 53; 55.
5	15; 17; *30; 47.
6	18; 46.
7	19.
8	20; 21; 22; 39.
9	23; 44.
10	24.
11	36; 40; 50; *51.
12	52.

* K4 is common to groups 03 and 04.
* K30 is common to groups 03 and 05.
* K33 is common to groups 01 and 03.
* K51 is common to groups 03 and 011.
** TNK11 is a proposed new K type.

with the results of both early and more recent, limited, feeding tests on human volunteers in which only Kanagawa positive strains induced illness (Sakazaki *et al.*, 1968; Sanyal & Sen, 1974). Although the reasons for these findings have still to be explained, and despite occasional outbreaks of food poisoning due to Kanagawa negative strains (Zen-Yoji *et al.*, 1973), this reaction does nevertheless seem to distinguish potentially virulent from less virulent strains in all countries where human infection with *V. parahaemolyticus* has been identified. Unfortunately, this test has been modified or incorrectly read by so many different workers that many of the results from different sources, especially those of marine isolates, are suspect and comparisons therefore of limited value. Indeed, at the International Symposium, the need for uniformity in performing this empirical but useful test was emphasized and precise instructions for carrying out and reading it are therefore given below.

(a) *The Kanagawa test*

(i) *Wagatsuma medium*

This medium contains: yeast extract, 5 g; Trypticase (BBL), 10 g; NaCl, 70 g; mannitol, 5 g; crystal violet, 1 mg; agar, 15 g; distilled water, 1 l; pH, 7.5.

The mixture of ingredients is heated carefully with frequent agitation until the agar is completely dissolved, but it is not autoclaved. The medium is cooled to 50° and 10 ml of a 20% suspension of washed human red blood cells (freshly obtained) are then added to each 100 ml of the molten medium. The medium is mixed thoroughly and poured into Petri dishes.

(ii) *Method of testing*

Plates of Wagatsuma medium are spot-inoculated with loopfuls of overnight broth cultures of control and test strains of *V. parahaemolyticus*. The plates are incubated at 37° and read after 18–24 h (not later). Positive cultures give clear zones of β-haemolysis. Discoloration and α-haemolysis should be regarded as negative. Stab inoculation of single colonies gives excellent results.

(b) *Significance of the Kanagawa reaction*

The physico-chemical nature of the factors concerned in the Kanagawa reaction has been investigated extensively and several components have been identified, including a heat-labile and a heat-stable haemolysin (Sakazaki, 1973a). Although probably not directly responsible for virulence, the latter does induce antibody formation in patients following infection and it is possibly restricted to pathogenic strains. However, the actual mechanism of pathogenicity, whether due to enterotoxin production, invasive properties or both, still requires elucidation. Despite considerable work on immunochemistry, transfer factors, and comparison of live cultures, filtrates, cell lysates and other preparations from Kanagawa positive as well as negative strains in various tests, including ligated ileal loops, germ-free mice and tissue cultures, the evidence so far is still uncertain (Twedt & Brown, 1973). One suggestion is that heat-stable haemolysin, Kantigen and enterotoxin, although separate entities, are closely associated with each other and may be plasmid-mediated; however, there is no clear epidemiological relation between serotype and infection. Sakazaki *et al.* (1974) suggest that the ability of Kanagawa positive strains to multiply more rapidly than Kanagawa negative strains in the gut may be the essential factor responsible for virulence, and that enterotoxic substances, regardless of the Kanagawa reaction, may contribute to their pathogenicity. This suggestion accords with experimental evidence obtained by Barrow & Miller (1974) from growth studies in digest broths prepared from various fish and crustacea. Crab and prawn broths, compared with meat broth, both gave luxuriant growth of *V. parahaemolyticus,* yielding 10^7–10^8 viable organisms/ml after overnight incubation at 37° (Fig. 1). During the log phase of growth in prawn broth the minimal generation time was estimated to be very short and of the order of 5–7 min. There was also suggestive evidence that Kanagawa positive strains might survive and grow more quickly than Kanagawa negative strains under certain conditions. In view of these results and in order to simulate roughly conditions in the alimentary tract after eating seafoods, acid prawn broth

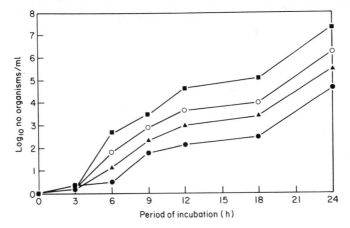

Fig. 1. Comparative growth of *Vibrio parahaemolyticus* in fish and meat digest broths at 37°. Digest broths: ■, prawn; ○, crab; ▲, cod; ●, meat.

(pH 6.0) was inoculated with approximately equal numbers of a Kanagawa positive and a Kanagawa negative strain; after incubation at 37° for 1 h, the pH value was adjusted to 8.0. As Fig. 2 shows, there was an initial decrease in their numbers, but within 3–6 h, growth of the Kanagawa positive organisms was better than and within 12 h greatly exceeded that of the Kanagawa negative strain. If this occurs naturally in the gut, it would certainly help to explain why Kanagawa positive strains are usually isolated from patients and are therefore regarded as more 'virulent'.

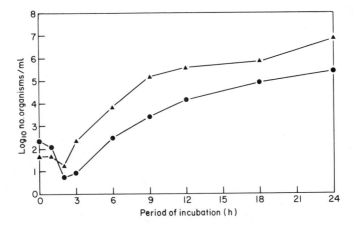

Fig. 2. Comparative growth of Kanagawa positive and negative strains of *Vibrio parahaemolyticus* in prawn broth at pH 6.0, adjusted after 1 h at 37° to pH 8.0. ▲, Kanagawa positive strain; ●, Kanagawa negative strain.

6. Epidemiology

Illness caused by *V. parahaemolyticus* is almost invariably foodborne, usually from seafoods, although occasional outbreaks due to cross contamination of other foods may occur (Kudoh *et al.,* 1974). Except for those eaten raw, the seafoods concerned have usually been contaminated after cooking, from other marine sources. Spread of infection from one person to another has not yet been recognized and the possible significance of carriers is doubtful. Although the infective dose needed to cause illness is probably large, the organism can multiply rapidly, even at room temperature (18–22°), so that only a short time is necessary for the production of hazardous numbers. Seafoods responsible for illness vary with season and local eating habits in different countries (Barrow, 1973*b*) but *V. parahaemolyticus* has now been isolated from oysters, clams, shrimps, prawns, crab and lobster meat and market fish as well as from other marine sources throughout the world. Outbreaks of food poisoning, all associated with seafoods, caused by this organism have recently been recognized in many of these countries (Battey *et al.,* 1970; Bockemühl, Amédomé & Triemer, 1972; Barker, 1974). The presence of *V. parahaemolyticus* has been confirmed in British coastal waters and seafoods (Barrow & Miller, 1969, 1972), and 2 outbreaks of infection with this organism have been recognized in Britain among persons who had eaten crab meat, one at a local holiday resort (Hooper, Barrow & McNab, 1974), and the other affecting airline passengers and crew returning from abroad (Table 4). In the airline outbreak (Peffers *et al.,* 1973), examination of numerous colonies from primary cultures of the crab meat in *hors d'oeuvres* revealed, for the first time, occasional Kanagawa positive colonies of the same serotype as that isolated from the patients. This actual food-patient correlation also suggests that the few Kanagawa positive strains present in the food grew rapidly and preferentially in those affected.

Water temperature is an important factor governing growth and survival of *V. parahaemolyticus* in the marine environment. Its numbers decline rapidly below 15°, and it can rarely be isolated except in small numbers from water or sediment below 10°. Barrow & Miller (1974) found that Kanagawa positive as well as Kanagawa negative strains both survived well below 25° in autoclaved seawater but at 37° the Kanagawa positive strains lost their ability to cause β-lysis in Wagatsuma medium and would thus be regarded as Kanagawa negative. Paradoxically, they retained this ability below 25°, although at this temperature, Kanagawa negative organisms were able to grow better. This experimental evidence was further supported by the isolation, during seafood examinations, of several strains which were Kanagawa positive on primary isolation at 37°, but which rapidly became negative after subculture. These isolations may be related in some way to greater buffering capacity due to the presence of crushed shells. Examination of multiple colonies from primary cultures of various seafoods

Table 4

Isolations of Vibrio parahaemolyticus *from Patients in Britain*

Occurrence of food poisoning	No. of cases known	Serotype	Kanagawa reaction	Food involved
Fatal case of acute gastroenteritis	1	03 : K4	+	British cockles
Outbreak of acute gastro-enteritis at holiday camp	12	01 : K56	+	British dressed crab
Outbreak of acute gastro-enteritis on aircraft from Bangkok	12	02 : K3	+	Thai dressed crab in *hors d'oeuvres*
Acute gastroenteritis return by air from Bangkok	1	02 : K3	−	Thai prawns
Mild diarrhoea after return by air from Bangkok	1	03 : K4	−	Not known
Diarrhoea after return by air from Bangkok	1	03 : K48	−	? prawns
Acute gastroenteritis after return by air from West Africa	1	08 : K22	+	West African shrimps
Restaurant barman who ate large amounts of imported prawns intended for customers	1	04 : K9	+	Imported prawns ? origin
Acute gastroenteritis after return from Australia by air	7	04 : K9	+	Prawn cocktail ? Malaysian prawns
Acute gastroenteritis after return by air from Far East	1	01 : K26	−	Not known

have, however, enabled the detection of occasional 'true' Kanagawa positive colonies, usually among multiple Kanagawa negative serotypes (Peffers *et al.*, 1973; Wagatsuma, 1974). These findings suggest that, irrespective of the Kanagawa reaction, serological typing of isolates from marine sources is important epidemiologically, and may indeed be more important than the Kanagawa reaction itself. This perhaps should be performed initially on primary isolates as well as after serial subculture.

Vibrio parahaemolyticus is probably present in warm and temperate coastal waters throughout the world, and it is therefore likely to be found, usually in small numbers, in most seafoods from such areas. It is not normally present in deep sea fish when caught, but they almost inevitably become contaminated during marketing. Excluding countries where marine products are customarily eaten raw, most seafoods are generally cooked shortly before consumption and thus do not present any real bacterial hazard. Commercially-produced crab-meat, however, is usually extracted by hand after cooking and it, thus, constitutes a greater hazard, especially as deep-freezing is not as lethal to *V. parahaemolyticus* as is chilling at 0–5° (Liston, 1974). Filter-feeding shellfish, especially oysters and mussels, which concentrate bacteria and other food particles physiologically from the large volume of water passed through their gills, are often sold live and eaten raw and are therefore potentially hazardous. However, the use of purification plants or relaying in sewage-free waters, together with bacteriological monitoring, have done much to ensure clean and safe supplies of shellfish. Since *V. parahaemolyticus* is fully sensitive to heat, as well as to acid-preservation, its control should theoretically present few difficulties, but in practice, the vast international trade in raw and cooked seafoods does pose some problems with certain imported products.

7. Imported Seafoods

The potential hazards from *V. parahaemolyticus* in imported seafoods vary with the country of origin. In some countries, the adequacy of cooking and hygiene during subsequent handling are probably the most important factors. Commercially, consignments of raw as well as cooked seafoods, particularly shrimps and prawns from the Far East, are imported by many countries in the Western Hemisphere for processing and repacking. This involves thawing bulk quantities for packaging and labelling in smaller units, ranging from domestic to catering size packs. These are refrozen for subsequent distribution, both at home and abroad — sometimes without any indication of their original source. Despite some reduction in bacterial numbers due to low temperature refrigeration, some of these crustacea may contain large numbers of marine bacteria, even in cooked products, thus denoting poor hygiene during processing. In tropical and perhaps less advanced countries, other factors such as water quality, ice-cleanliness, preparation techniques, sanitation and working conditions add to the difficulties of producing hygienic products. Advice, field assistance and on-site monitoring of processing by importers is therefore important. As an alternative, perhaps such seafoods should only be imported raw so that the final processing can be adequately controlled.

8. Safety

Whatever criteria for quality may be set by importers or by countries (Shewan, 1970), there seems little point in using *V. parahaemolyticus* itself as a yardstick for safety since it is almost invariably outnumbered by other marine bacteria. It would seem more appropriate to use the common marine organism *V. alginolyticus* or a total viable count on a seawater medium, as well as enteric organisms and *Staphylococcus aureus,* as additional indicators of satisfactory processing. Such standards would presumably be applied equally to processed seafoods for retail, whether home-produced or imported. Equally important, but undoubtedly more difficult, are the questions as to what constitutes a representative sample and what action, if any, should follow the results of bacteriological tests. Despite the cases and outbreaks so far identified, there is, as yet, relatively little evidence to suggest that in Britain, *V. parahaemolyticus* represents a significant health hazard in either home-produced or imported seafoods. Although it is important to improve and maintain hygienic standards in seafoods, and despite the ability of this organism to multiply rapidly, it is equally important not to condemn otherwise sound and perhaps expensive food without good cause. Exporters, however, may have to meet current trends which seem to favour more rigid numerical standards, so that uniformity of techniques is clearly important.

If people choose to eat raw or undercooked foods, they expose themselves, whether knowingly or not, to certain unnecessary risks, and it cannot be emphasized too strongly that safety ultimately depends, not so much on arbitrary microbiological standards, as on hygienic production, correct storage and good distribution together with education in intelligent eating habits. With the current worldwide emphasis on intensive fish farming, both in natural waters and in warm effluents from power plants, *V. parahaemolyticus* may well become a greater hazard in the future.

9. References

ANON. (1974). International Symposium on *Vibrio parahaemolyticus,* Eds T. Fujino, G. Sakaguchi, R. Sakazaki & Y. Takeda. Tokyo: Saikon Publishing Co.

BAIN, N. & SHEWAN, J. M. (1968). Identification of *Aeromonas, Vibrio* and related organisms. In *Identification Methods for Microbiologists.* B. Eds B. M. Gibbs & D. A. Shapton. London: Academic Press.

BARKER, W. H. (1974). *Vibrio parahaemolyticus* outbreaks in the United States. *Lancet,* 1, 551.

BAROSS, J. & LISTON, J. (1968). Isolation of *Vibrio parahaemolyticus* from the North-West Pacific. *Nature, Lond.* 217, 1263.

BARROW, G. I. (1973a). The environmental significance of *Vibrio parahaemolyticus* and other marine vibrios. In Proceedings of the 6th International Symposium of the World Association of Veterinary Food Hygienists, Elsinore, Denmark.

BARROW, G. I. (1973b). Marine micro-organisms and food poisoning. In *The Microbiological Safety of Food.* Eds B. C. Hobbs & J. H. B. Christian. London: Academic Press.

BARROW, G. I. & MILLER, D. C. (1969). Marine bacteria in oysters purified for human consumption. *Lancet,* 2, 421.

BARROW, G. I. & MILLER, D. C. (1972). *Vibrio parahaemolyticus:* a potential pathogen from marine sources in Britain. *Lancet,* 1, 485.

BARROW, G. I. & MILLER, D. C. (1974). Growth studies on *Vibrio parahaemolyticus* in relation to pathogenicity. In Proceedings of the International Symposium on *Vibrio parahaemolyticus* (see Anon. 1974).

BATTEY, Y. M., WALLACE, R. B., ALLAN, B. C. & KEEFFE, B. M. (1970). Gastroenteritis in Australia caused by a marine vibrio. *Med. J. Australia,* 1, 430.

BAUMANN, P., BAUMANN, L. & MANDEL, M. (1971). Taxonomy of marine bacteria: the genus *Beneckea. J. Bact.* 107, 268.

BERGEY'S MANUAL OF DETERMINATIVE BACTERIOLOGY (1974). 8th Edition. Eds R. E. Buchanan & N. E. Gibbons. Baltimore: The Williams and Wilkins Co.

BOCKEMÜHL, J., AMÉDOMÉ, A. & TRIEMER, A. (1972). Gastro-entérites cholériformes dues à *Vibrio parahaemolyticus* sur la côte du Togo (Afrique occidentale). *Z. Tropenmed. Parasit.* 23, 308.

CHATERJEE, B. D. (1974). Epidemiologic and taxonomic status of *Vibrio parahaemolyticus.* In Proceedings of the International Symposium on *Vibrio parahaemolyticus* (see Anon. 1974).

FUJINO, T., OKUNO, Y., NAKADA, D., AOYAMA, A. FUKAI, K., MUKAI, T. & UEHO, T. (1953). On the bacteriological examination of shirasu food-poisoning. *Med. J. Osaka Univ.* **4,** 299.

HOOPER, W. L., BARROW, G. I. & McNAB, D. J. N. (1974). *Vibrio parahaemolyticus* food-poisoning in Britain. *Lancet,* **1,** 1100.

HUGH, R. & SAKAZAKI, R. (1972). Minimal number of characters for the identification of *Vibrio* species, *Vibrio cholerae,* and *Vibrio parahaemolyticus. Pub. Hlth. Lab.* **30,** 133.

KANEKO, T. & COLWELL, R. R. (1973). Ecology of *Vibrio parahaemolyticus* in Chesapeake Bay. *J. Bact.* **113,** 24.

KRISTENSEN, K. K. (1974). Semi-quantitative examinations on the contents of *Vibrio parahaemolyticus* in the sound between Sweden and Denmark. In Proceedings of the International Symposium on *Vibrio parahaemolyticus* (see Anon. 1974).

KUDOH, Y., SAKAI, S., ZEN-YOJI, H. & LeCLAIR, R. A. (1974). Epidemiology of food-poisoning due to *Vibrio parahaemolyticus* occurring in Tokyo during the last decade. In International Symposium on *Vibrio parahaemolyticus* (see Anon. 1974).

LISTON, J. (1973). *Vibrio parahaemolyticus.* In *Microbial Safety of Fishery Products.* Eds C. O. Chichester & H. D. Graham. New York: Academic Press.

LISTON, J. (1974). Influence of U.S. Seafood Handling Procedures on *Vibrio parahaemolyticus.* In Proceedings of the International Symposium on *Vibrio parahaemolyticus* (see Anon. 1974).

LORENZO, F., SOSCIA, M., MANZILLO, G. & BALESTRIERI, G. G. (1974). Epidemic of cholera El Tor in Naples, 1973. *Lancet,* **1,** 669.

MIYAMOTO, Y., KATO, T., OBARA, Y., AKIYAMA, S. & TAKIZAWA, K. (1969). In vitro hemolytic characteristic of *Vibrio parahaemolyticus:* its close correlation with human pathogenicity. *J. Bact.* **100,** 1147.

MIYAMOTO, Y., NAKAMURA, K. & TAKIZAWA, K. (1961). Pathogenic halophiles. Proposal of a new genus *'Oceanomonas'* and of the amended species names. *Jap. J. Microbiol.* **5,** 477.

MONSUR, K. A. (1963). Bacteriological diagnosis of cholera under field conditions. *Bull. Wld. Hlth. Org.* **28,** 387.

PEFFERS, A. S. R., BAILEY, J., BARROW, G. I. & HOBBS, B. C. (1973). *Vibrio parahaemolyticus* gastroenteritis and International Air Travel. *Lancet,* **1,** 143.

ROLAND, F. P. (1970). Leg gangrene and endotoxin shock due to *Vibrio parahaemolyticus* – an infection acquired in New England coastal waters. *New Eng. J. Med.* **282,** 1306.

SAKAI, S., KUDOH, Y. & ZEN-YOJI, H. (1974). Food-poisoning caused by indole negative strains of *Vibrio parahaemolyticus.* In Proceedings of the International Symposium on *Vibrio parahaemolyticus* (see Anon. 1974).

SAKAZAKI, R. (1968). Proposal of *Vibrio alginolyticus* for the biotype 2 of *Vibrio parahaemolyticus. Jap. J. Med. Sci.* **21,** 359.

SAKAZAKI, R. (1969). Halophilic Vibrio Infections: In *Food-Borne Infections and Intoxications.* Ed. H. Riemann. New York: Academic Press.

SAKAZAKI, R. (1973*a*). Recent Trends of *Vibrio parahaemolyticus* as a causative agent of food-poisoning. In *The Microbiological Safety of Food.* Eds B. C. Hobbs & J. H. B. Christian. London: Academic Press.

SAKAZAKI, R. (1973*b*). Control of Contamination with *Vibrio parahaemolyticus* in seafoods and Isolation and Identification of the Vibrio. In *The Microbiological Safety of Food.* Eds B. C. Hobbs & J. H. B. Christian. London: Academic Press.

SAKAZAKI, R., IWANAMI, S. & FUKUMI, H. (1963). Studies on the enteropathogenic, facultatively halophilic bacteria, *Vibrio parahaemolyticus.* I. Morphological cultural and biochemical properties and its taxonomical position. *Jap. J. Med. Sci. Biol.* **16,** 161.

SAKAZAKI, R., IWANAMI, S. & TAMURA, K. (1968). Studies on the enteropathogenic facultatively halophilic bacteria., *Vibrio parahaemolyticus*. II. Serological characteristics. *Jap. J. Med. Sci. Biol.* **21**, 313.

SAKAZAKI, R., TAMURA, K., KATO, T., OBARA, Y., YAMAI, S. & HOBO, K. (1968). Studies on the enteropathogenic facultatively halophilic bacteria, *Vibrio parahaemolyticus*. III. Enteropathogenicity. *Jap. J. Med. Sci. Biol.* **21**, 325.

SAKAZAKI, R., TAMURA, K., NAKAMURA, A., KURATA, T., GOHDA, A. & KAZUNO, Y. (1974). Studies on enteropathogenicity of *Vibrio parahaemolyticus* using ligated gut loop model in rabbits. *Jap. J. Med. Sci. Biol.* **27**, 35.

SANYAL, S. C. & SEN, P. C. (1974). Human volunteer study on the pathogenicity of *Vibrio parahaemolyticus*. In Proceedings of the International Symposium on *Vibrio parahaemolyticus* (see Anon. 1974).

SHEWAN, J. M. (1970). Bacteriological Standards for fish and fishery products. *Chemistry and Industry*, p. 193.

TAKIKAWA, I. (1958). Studies on pathogenic halophilic bacteria. *Yokohama Med. Bull.* **2**, 313.

TAMURA, K., SHIMADA, S. & PRESCOTT, L. M. (1971). Vibrio agar: a new plating medium for isolation of *V. cholerae*. *Jap. J. Med. Sci. Biol.* **24**, 125.

TWEDT, R. M. & BROWN, D. F. (1973). *Vibrio parahaemolyticus:* Infection or Toxicosis? *J. Milk Food Technol.* **36**, 129.

WAGATSUMA, S. (1974). Ecological studies on Kanagawa phenomenon positive strains of *Vibrio parahaemolyticus*. In Proceedings of the International Symposium on *Vibrio parahaemolyticus* (see INTERNATIONAL SYMPOSIUM, 1974).

YABUUCHI, E., MIWATANI, T., TAKEDA, Y. & ARITA, M. (1974). Flagellar Morphology of *Vibrio parahaemolyticus*. In Proceedings of the International Symposium on *Vibrio parahaemolyticus* (see Anon. 1974).

ZEN-YOJI, H., LeCLAIR, R. A., OHTA, K. & MONTAGUE, T. S. (1973). Comparison of *Vibrio parahaemolyticus* cultures isolated in the United States with those isolated in Japan. *J. Infect. Dis.* **127**, 237.

Bacillus cereus Food Poisoning

R. J. Gilbert and A. J. Taylor*

Food Hygiene Laboratory, Central Public Health Laboratory, Colindale Avenue, London NW9 5HT, England

CONTENTS

1. Introduction

PAPERS AND REVIEWS on the microbiological aspects of food poisoning are frequently confined to a consideration of *Salmonella* spp., *Staphylococcus aureus, Clostridium welchii, Cl. botulinum* and more recently *Vibrio parahaemolyticus*. While these organisms are without doubt the most commonly reported agents of food poisoning, other bacteria have from time to time been implicated.

Since the turn of the century several accounts of food poisoning associated with aerobic spore-forming bacilli have been reported in the European literature (Lubenau, 1906; Seitz, 1913; Brekenfeld, 1926, 1929; Trüb & Wundram, 1942; Plazikowski, 1949). These reports shared several features: they rarely presented a complete description of the implicated organism, but usually classified it as an 'anthracoid' or 'pseudoanthrax' bacillus or as a member of the 'subtilis-mentericus' group, and did not give the number of these and other organisms in the incriminated foods.

Since 1950, however, and particularly in recent years, there has been an increasing number of well documented reports which have established *Bacillus cereus* as a food poisoning organism. These reports may represent a genuine increase in the number of incidents, but more likely they indicate a growing awareness of the problem in many countries. Goepfert, Spira & Kim (1972) have reviewed in considerable detail the evidence pertinent to the role of *B. cereus* as a food poisoning agent together with information on the properties of the organism and methods for its isolation, identification and enumeration.

* Present address: Diagnostic Bacteriology Laboratory, St. Mary's Hospital Medical School (University of London), Norfolk Place, London W2.

This paper aims to review, briefly, the state of knowledge on *B. cereus* food poisoning, to describe outbreaks attributed to this organism which have occurred recently in this and other countries and which differ in several respects from those hitherto reported, and to indicate areas where more information and research are required.

2. *Bacillus cereus* and its Isolation

Bacillus cereus was first described by Frankland & Frankland (1887). It belongs to the morphological Group 1 of Smith, Gordon & Clark (1952) and is a large celled organism, 1.0–1.2 μm x 3.0–5.0 μm, sporing readily on nutrient media. Smith *et al.* (1952) subdivided *B. cereus* into 4 variants — *B. cereus* var. *cereus*; var. *thuringiensis*; var. *anthracis*; and var. *mycoides,* but we will follow the example of Goepfert *et al.* (1972) and retain species rank for the 4 variants.

Bacillus cereus is common in soil and on vegetation and has been isolated in several countries from a wide variety of foods (Nygren, 1962; Mossel, Koopman & Jongerius, 1967; Kim & Goepfert, 1971*a*). It is responsible for a spoilage problem referred to as 'broken' or 'bitty' cream (Stone & Rowlands, 1952; Stone, 1952*a,b*; Davies & Wilkinson, 1973).

The organism grows in the temperature range 10°–48° with an optimum between 28° and 35° (Smith *et al.,* 1952). Colony morphology on horse blood agar is characteristic. Incubation at 35°–37° for 48 h produces large, 4–7 mm diam., flat, matt colonies tending towards a green coloration. Isolates are usually α-haemolytic, but some strains produce β-haemolysis. Under anaerobic incubation the colonies are small and translucent, 2–3 mm in diameter, surrounded by a zone of β-haemolysis and resembling colonies of *Cl. welchii* but usually possessing an irregular edge.

Interest in the isolation of *B. cereus* from foodstuffs has led to the formulation of a number of selective and/or differential media and the use of some of these has been reviewed in detail by Inal (1971, 1972). Hauge (1955) used blood agar incubated first anaerobically for 18 h then aerobically for a further 18 h in order to differentiate *B. cereus* from *Cl. welchii.*

Donovan (1958) used a peptone-beef extract-egg yolk agar containing lithium chloride and polymyxin B as selective agents: typical colonies of *B. cereus* were surrounded by an opaque zone after 18 h at 30°. Mossel *et al.* (1967) introduced mannitol and phenol red into an egg yolk-polymyxin medium in order to improve differentiation and Kim & Goepfert (1971*b*) described another egg yolk-polymyxin medium which enhances sporulation and allows serological identification with the use of species-specific antisera.

All of these media may be used for the enumeration of *B. cereus* in foodstuffs. They rely upon the suppression of Gram negative organisms by polymyxin and the presumptive identification of *B. cereus* by means of the egg yolk reaction. It is therefore of interest that in 2 of the outbreaks in this country in 1973 the strains of *B. cereus* isolated gave a feeble egg yolk reaction. In contrast, the colonies on blood agar were characteristic of *B. cereus.*

It is our experience that the isolation and enumeration of *B. cereus* in foodstuffs is adequate on horse blood agar incubated aerobically at 35–37° for 48 h. This method has proved successful for performing colony counts on faecal specimens and grossly contaminated foodstuffs.

Nygren (1962) isolated small numbers of the organism from up to 52% of a variety of foodstuffs using a Most Probable Number technique, but the levels of *B. cereus* in foodstuffs incriminated in food poisoning outbreaks are sufficiently high to make enrichment not only unnecessary but misleading. When present in small numbers, *B. cereus* may be isolated by incubation of the food in nutrient broth at 37° for 18 h with subsequent subculture on blood agar.

Confirmation of isolates as *B. cereus* is made by sub-culture on Kendall's B.C. medium and on ammonium salt-sugar media containing glucose, arabinose, mannitol or xylose (Cowan & Steel, 1965).

(a) *Preparation and use of B.C. medium*

B.C. medium is made by adding mannitol and egg yolk emulsion to a basal medium, as described below (Miss M. Kendall, pers. comm.). The basal medium (Cowan & Steel, 1965) contains: $(NH_4)_2HPO_4$, 1 g; KCl, 0.2 g; $MgSO_4 7H_2O$, 0.2 g; Yeastrel, 0.2 g; agar, 20 g; distilled water, 1 l.

The ingredients are mixed, steamed to dissolve, filtered, the pH value adjusted to 7.0 and 4 ml of 1% alcoholic bromocresol purple added. The base is distributed in 90 ml amounts and autoclaved at 115° for 10 min.

Eggs are separated aseptically and the yolks broken and mixed with an equal volume of physiological saline, filtered through gauze, distributed into sterile bottles and heated at 60° for 30 min. The emulsion is stored at 4° until required.

To 90 ml of basal medium, 1 g of mannitol is added. The base is melted by heating at 115° for 10 min. and allowed to cool to 50° before adding 10 ml of egg yolk emulsion. After mixing well, plates are poured, allowed to set and dried at 37° for 1–2 h.

The medium is inoculated by stabbing with a straight wire or a loop. After incubation at 35°–37° for 18 h the bacterial growth is surrounded by an opaque zone (the egg yolk reaction) with purple coloration of the surrounding agar. Any organism which gives this typical appearance on B.C. medium and ferments only glucose is considered to be *B. cereus.*

3. Food Poisoning

Hauge (1950, 1955) published the first complete account of *B. cereus* food poisoning in his investigation of 4 outbreaks in Norway involving 600 persons. The food implicated in all 4 outbreaks was vanilla sauce prepared and stored at room temperature for one day before being served. Although samples of the sauce contained 25–110 x 10^6 B. cereus/ml, there was little change in the odour, taste or consistency of the product. Corn starch, which was one of the ingredients of the vanilla sauce powder, contained up to 10^4 B. cereus spores/g.

The average incubation period for the outbreaks in Norway was c. 12 h with a range of 8–16 h. The patients suffered abdominal pain, profuse watery diarrhoea, rectal tenesmus and moderate nausea which seldom resulted in vomiting. Fever was not a common symptom. The illness did not usually last > 12 h.

Since 1950, outbreaks of B. cereus food poisoning have been reported from several other European countries including Denmark, Italy, the Netherlands, Hungary, Sweden, Rumania, Germany and the U.S.S.R. The first well-documented outbreaks in the U.S.A. occurred in 1969 (Midura et al., 1970) and in Great Britain in 1971 (Anon. 1972).

An unusually wide range of foods have been implicated including meat and vegetable soups, cooked meat and poultry, cooked vegetables, fried rice, dessert dishes and occasionally foods such as fish, pasta, ice-cream and milk. Table 1 summarizes some of the published information from several countries together with data, where available, on the numbers of B. cereus in the incriminated foods.

It is noteworthy that B. cereus has not usually been isolated in large numbers from faecal specimens from patients. Hauge (1950), for example, described the growth of B. cereus from faecal specimens as 'usually very scarce'.

Hungarian investigators, particularly Nikodemucz and his colleagues, have studied various aspects of B. cereus food poisoning in some detail (Nikodemucz, 1958, 1965, 1966a,b, 1967, 1968; Nikodemucz et al., 1962, 1967, 1969). This work can be linked to the fact that B. cereus was the third most common cause of bacterial food poisoning in Hungary during the period 1960 to 1968 being responsible for c. 8% of reported outbreaks and c. 15% of reported cases (Ormay & Novotny, 1970, see Table 2). Over half the outbreaks involved meat or meat products and Ormay & Novotny (1969) have attributed this to the fact that Hungarian meat dishes are frequently well seasoned with spices which often contain large numbers of aerobic spore-forming bacilli. Cooking of the food was usually insufficient to kill the spores and post-cooking storage conditions allowed surviving spores to germinate and the resulting vegetative cells to multiply.

The isolation of large numbers of B. cereus from foods implicated in food poisoning is strong presumptive evidence that the organisms were the agents responsible for illness. However, enteropathogenicity can only be proven conclusively by eliciting similar symptoms in human volunteers fed pure cultures of the organism.

Hauge (1955) consumed 200 ml of a vanilla sauce containing 92 x 10^6 B.

Table 1

Examples of food poisoning outbreaks attributed to Bacillus cereus

Reference	Country	Type of food	Count of B. cereus/g or /ml.
Hauge (1950, 1955)	Norway	Vanilla sauce	$2.5 \times 10^7 - 1.1 \times 10^8$ (4 outbreaks)
Christiansen, Koch & Madelung (1951)	Denmark	Yellow pudding dessert	1.3×10^7
Pisu & Stazzi (1952)	Italy	Chicken soup	6×10^7
Clarenburg & Kampelmacher (1957)	Netherlands	Mashed potatoes, vegetables, minced meat, liver sausage, Indonesian rice dishes, puddings, soups	$5 \times 10^5 - 2 \times 10^8$ (11 outbreaks)
Nikodemucz (1958)	Hungary	Vegetable soup	Not recorded
Kiss (1961)	Hungary	Sausage	10^5
Nikodemucz et al. (1962)	Hungary	Sausage, vegetable dishes, cream pastries, soups	$3.6 \times 10^4 - 9.5 \times 10^8$ (35 outbreaks)
Heinertz (1962)	Sweden	Pork casserole	2×10^7
Nygren (1962)	Sweden	Vanilla cream	9.2×10^5
Birzu et al. (1968)	Rumania	Cooked meat	$10^7 - 10^8$
Ormay & Novotny (1969)	Hungary	Meat (various), dishes with rice and forcemeat, vegetables, Italian pasta, milk	Not recorded
Midura et al. (1970)	U.S.A.	Meat loaf	7×10^7
Pivovarov et al. (1970)	U.S.S.R.	Sausage, borsch soup, vegetable soup, cooked meats, cooked vegetables, pilchards	Not recorded
Von Ludwig (1971)	Germany	Vanilla sauce	10^7
Vlad & Vlad (1972)	Rumania	Milk	2.1×10^7
Bulyba, Kul'chskaya & Domanskaya (1973)	U.S.S.R.	Ice-cream	Not recorded
Todd et al. (1974)	Canada	Cooked chicken	$6 \times 10^4 - 3 \times 10^8$ (2 outbreaks)
Anon. (1972, 1973, unpublished data) Mortimer & McCann (1974)	Great Britain	Fried and boiled rice, rice risotto, beef curry	$3 \times 10^5 - 2 \times 10^9$ (17 outbreaks)*
Lefebvre et al. (1973)	Canada	Fried rice	2.4×10^7
Taplin (1973, 1974, pers. comm.)	Australia	Fried rice	Not recorded

* See Table 4.

Table 2

Bacterial food poisoning of known aetiology in Hungary 1960–1968*

Causal agent	Outbreaks		Cases	
	Number	%	Number	%
Staphylococcus aureus	866	58.1	9 221	35.2
Salmonella	339	22.7	8 916	34.0
Aerobic sporeforming				
bacilli (*Bacillus cereus*)	125	8.4	3 871	14.8
Clostridium welchii	28	1.9	1 892	7.2
Other micro-organisms†	133	8.9	2 321	8.8
Totals	1 491	100	26 221	100

* Data from Ormay & Novotny (1970).
† *Pseudomonas aeruginosa, Proteus, Streptococcus faecalis, Klebsiella, Shigella, E. coli* and *Clostridium botulinum.*

cereus cells/ml. Thirteen hours later he experienced severe abdominal pain, diarrhoea and rectal tenesmus; the symptoms persisted for 8 h. In a previous experiment (Hauge, 1950) 6 volunteers drank 155–270 ml of vanilla sauce containing 30–60 x 10^6 *B. cereus*/ml and 4 developed typical symptoms. Dack *et al.* (1954) were unable to confirm Hauge's findings but the 4 cultures of *B. cereus* used were isolated from cheese and had not been associated with food poisoning. Johannesen (1957) and Nikodemucz *et al.* (1969) reported only partial success in volunteer tests.

Experiments in carnivores are similarly confusing. Nikodemucz (1965, 1966a,b, 1967) reported that cats and dogs fed food containing large numbers of *B. cereus* developed diarrhoea: the incubation period in experiments using a large number of dogs was between $2\frac{1}{2}$ and 7 h. In contrast, Chastain & Harris (1974) have described symptoms of vomiting and diarrhoea and vomiting and depression in 2 dogs suffering from food poisoning. Symptoms in the second dog began 4–5 h after ingestion of the contents of an open can of refrigerated dog food; large numbers of *B. cereus* were isolated in pure culture from the food.

In summary, *B. cereus* has been described as the causative agent of a type of food poisoning almost identical in both incubation period and symptoms to that caused by *Cl. welchii.* With few exceptions most reports have come from Europe and a wide variety of foods have been incriminated.

Recent outbreaks associated with cooked rice

Since 1971 at least 34 incidents of food poisoning attributed to *B. cereus* (Table 3) have been reported in Great Britain (Public Health Laboratory Service, 1972, 1973, unpublished information; Mortimer & McCann, 1974). Each

Table 3

Food poisoning attributed to Bacillus cereus *in Great Britain* (1971–1973)

Year	Number of reported incidents*	Number of reported cases	References
1971	6	15	Anon. (1972)
1972	5	21	Anon. (1973)
1973	23	79	Anon. (1973) and unpublished information

* General outbreaks, family outbreaks and sporadic cases.

episode has been characterized by an acute attack of nausea and vomiting usually between 1 and 5 h after a meal; in a small number of patients the incubation period has varied from 15 min to 11 h. Diarrhoea has not been a common feature, being reported in only *c.* 25% of patients.

Fried or boiled rice has been the only food common to all those affected. Thirty-two of the 34 incidents were associated with cooked rice (usually fried) from Chinese restaurants or 'take-away' shops and in the other two incidents beef curry served in a public house and a rice risotto dish served in a 'health food' restaurant were implicated.

Bacteriological examination of food remnants and faecal and/or vomit specimens has failed to yield any of the organisms usually associated with food poisoning. In most of the incidents large numbers of aerobic spore-forming bacilli, identified as *B. cereus,* have been isolated from remnants of cooked rice, clinical specimens or both. Plate counts on blood agar of *B. cereus* in samples of fried or boiled rice from 17 incidents have ranged from $3 \times 10^5 - 2 \times 10^9$/g with a median value of 5×10^7/g (Table 4). From the first 18 incidents *B. cereus* was isolated from 32 of 40 faecal specimens in numbers as high as 3×10^9/g. Most of

Table 4

Counts of Bacillus cereus *in fried or boiled rice from* 17 *incidents of food poisoning*

Count of *B. cereus*/g	Number of incidents	% of incidents
$< 10^5$	0	–
$10^5 - 9.9 \times 10^5$	1	6
$10^6 - 9.9 \times 10^6$	3	18
$10^7 - 9.9 \times 10^7$	6	35
$10^8 - 9.9 \times 10^8$	5	29
$> 10^9$	2	12

Range $3 \times 10^5 - 2 \times 10^9$/g: median 5×10^7/g.

the eight persons from whom the organism was not isolated submitted their specimens for examination several days after cessation of symptoms. Twenty-nine of the 34 incidents occurred during the summer months of June to September and only one between December and February.

The incidents involving Chinese restaurants and 'take-away' shops are associated with the practice of saving portions of boiled rice from bulk cooking until required for frying. The boiled rice is allowed to 'dry off' at ambient temperatures for varying periods of time from a few hours up to 3 days, but usually overnight. When required, the rice is either re-heated or, and more usually, it is fried quickly with beaten egg and a small quantity of oil. The beaten egg is not always freshly prepared and may itself contain large numbers of various microorganisms. Thereafter, the fried rice is kept warm until served or is stored at ambient temperature and 'flash' fried again before being served. The situation is made worse by the preparation of large bulks of boiled rice which take several hours to cool down.

There is a reluctance among Chinese restaurateurs to store boiled rice in a refrigerator because they say that the grains of rice stick together, and it becomes difficult to 'toss' them in beaten egg during frying. However, recent work in this laboratory (Gilbert, Stringer & Peace, 1974) has shown that the storage of boiled or fried rice at kitchen temperature provides excellent conditions for the germination and outgrowth of *B. cereus* spores which have survived the boiling and/or frying process.

The slow cooling and non-refrigerated storage of cooked rice, indeed of all cooked foods, provide ideal conditions for bacterial growth particularly from surviving spores. To prevent further outbreaks of food poisoning from cooked rice Gilbert *et al.* (1974) have suggested that:

1. Rice should be boiled in smaller quantities on several occasions during the day, thereby reducing the storage time before frying.
2. After boiling, the rice should either be kept hot or cooled quickly and transferred to a refrigerator within 2 h of cooking. The cooling of boiled rice, especially large bulks, will be hastened by dividing the product into separate portions or by spreading the bulk in clean shallow containers.
3. Boiled or fried rice must not be stored under warm conditions and never at a temperature between $15°–50°$. Under no circumstances, therefore, should cooked rice be stored at kitchen temperature for > 2 h.
4. The beaten egg used in the preparation of fried rice should be freshly prepared.

Similar outbreaks with respect to incubation period, symptoms, food vehicles and the isolation of large numbers of *B. cereus* have been reported from Australia (Taplin, pers. comms.) and Canada (Lefebvre *et al.,* 1973). The incubation period and symptoms of outbreaks in Great Britain, Australia and

Table 5

Food poisoning caused by Clostridium welchii, Bacillus cereus and Staphylococcus aureus:
some clinical and epidemiological data

	Cl. welchii	B. cereus*	B. cereus†	Staph. aureus
Incubation period (h)	8–22	8–16	1–5	2–6
Duration of illness (h)	12–24	12–24	6–24	6–24
Diarrhoea	Extremely common	Extremely common	Fairly common	Common
Vomiting	Rare	Occasional	Extremely common	Extremely common
Foods most frequently implicated	Cooked meat and poultry	Meat products, soups, vegetables, puddings and sauces	Fried rice from Chinese restaurants and 'take-away' shops	Cooked meat and poultry and dairy products

* Outbreaks reported since 1950 in several countries including Norway, Denmark, Italy, The Netherlands, Hungary, Sweden, Rumania, Germany, The U.S.S.R., The U.S.A. and Canada.
† Outbreaks reported since 1971 in Great Britain: similar incidents have been reported in Australia and Canada.

Canada closely resemble those for staphylococcal food poisoning (Table 5). However, *Staph. aureus* has not been isolated from foods or clinical specimens in any of these outbreaks and staphylococcal enterotoxin was not detected in 3 samples of cooked rice from separate outbreaks in this country. In contrast, the outbreaks of *B. cereus* food poisoning in many other countries resemble *Cl. welchii* food poisoning with respect to incubation period and symptoms (Table 5).

4. Biochemical Variants and Serology

Extensive studies of biochemical characteristics were made on strains of *B. cereus* isolated in our laboratory and on cultures sent from a number of Public Health Laboratory Service and Hospital Laboratories in order to recognize (1) biochemical similarities between strains of *B. cereus* from food and faeces and/or vomit specimens from the same outbreak and (2) common features

Table 6
Biochemical characteristics of Bacillus cereus *strains*

Test	Result	% positive*	
Indole	–	0	
Voges – Proskauer	+	98.6	
Citrate utilization (Christensen)	+	93.1	
Nitrate reduction	+	93.7	
Urease (Christensen)	–	11.9	
Gelatin liquefaction	+	100	
Casein hydrolysis	+	100	
Potato starch hydrolysis	+	99.6	
Anaerobic growth in 1% glucose broth	+	100	
Egg yolk turbidity	+	100	
Motility	+	97.9	
Acid from: Glucose	+	100	
Mannitol	–	0	
Lactose	–	8.5	(113 strains)
Sucrose	+	89.8	(113 strains)
Maltose	+	100	(113 strains)
Inositol	–	0	(113 strains)
Dulcitol	–	0	(113 strains)
Xylose	–	0	
Glycerol	+	98.3	(113 strains)
Arabinose	–	0	
Trehalose	+	100	(113 strains)
Salicin	v	63.7	(113 strains)
Sorbitol	–	0	(113 strains)

* ++, positive in 80–100% of strains tested.
  –, negative in 80–100% of strains tested.
  v, variable reaction.
  461 strains examined except where indicated.

among all the isolates from outbreaks which would distinguish such strains from many of those isolated from routine foods and environmental sources. The biochemical characteristics of 461 strains of B. cereus are summarized in Table 6.

In 3 outbreaks we received foods or cultures which yielded isolates of B. cereus possessing distinguishing biochemical features. In 2 of these, strains from food and faeces were received and the results of the appropriate tests are shown in Table 7.

Burdon (1956) suggested that salicin fermentation was one means of distinguishing B. cereus from B. anthracis. The results of the salicin test performed on 113 of the cultures in our collection are shown in Table 8. All 113

Table 7
Biochemical variants among isolates of Bacillus cereus *from two outbreaks of food poisoning*

Outbreak A

Source of culture	Nitrate reduction	Citrate utilization	Egg-yolk reaction
Boiled rice	−	−	Narrow zone
Fried rice	−	−	Narrow zone
Beef curry	+	+	Wide zone
Faeces (1 patient)	−	−	Narrow zone

Outbreak B

Source of culture	Urease	Acid from salicin
Uncooked rice	+	−
Fried rice	+	−
Sweet and sour sauce	+	−
Bean sprouts	−	+
Faeces (4 patients)	+	−

Table 8
Salicin fermentation reactions among isolates of Bacillus cereus *

Isolates from	Number of cultures tested	Number positive	% positive
Routine foods	90	72	81.1
Food poisoning strains from Great Britain and Australia	23	0	0

* One representative strain, usually a faecal isolate from each outbreak.

strains grew at $45°$, rapidly liquefied gelatin at $22°$ and all but 2 were resistant to penicillin (tested by a disc sensitivity method using discs containing 10 i.u. of penicillin G.); these are features of *B. cereus* rather than of *B. anthracis*. In contrast, all the strains incriminated in the rice food-poisoning outbreaks failed to ferment salicin, whilst 81% of strains isolated from a wide range of foods by enrichment culture did ferment this glucoside. Strains in our collection from European and American outbreaks all ferment salicin as do the isolates described in the literature by Midura *et al.* (1970) and Todd *et al.* (1974). The relationship, if any, of salicin fermentation to toxin production is unknown.

The serological investigation of the outbreak strains is at an early stage. Norris & Wolf (1961) studied the spore, somatic and flagellar antigens of *B. cereus* in order to assess the taxonomic value of the different types of antigen. Their findings suggested that whilst the spore antigen possessed the highest species specificity, the flagellar 'H' antigen showed the best strain specificity.

Rabbits have been used to produce agglutinating sera against the 'H' antigens of strains of *B. cereus* from a number of outbreaks. Titres were in the range 1280–20 480 and preliminary results indicated that a few specific serotypes are involved in most outbreaks associated with the consumption of cooked rice (Taylor & Gilbert, 1975).

5. Pathogenicity

The relatively sudden onset of symptoms especially in outbreaks in this country and the short duration of illness indicate that *B. cereus* food poisoning is an intoxication rather than an infection. Nygren (1962) suggested that the symptoms of both *Cl. welchii* and *B. cereus* food poisoning were due to the liberation of phosphorylcholine from lecithin by the phospholipase C enzyme produced by these organisms. Duncan & Strong (1971) and Hauschild, Niilo & Dorward (1971), however, have shown that the symptoms of *Cl. welchii* food poisoning are due to a protein enterotoxin produced in the intestine by sporulating cells of toxigenic strains and which is separate and distinct from phospholipase C.

Goepfert (1973) and Goepfert *et al.* (1973) have shown that *B. cereus* also produces an enterotoxic substance that can be readily differentiated from lecithinase. The toxic substance is described as 'protein in nature, actively synthesized and secreted during logarithmic growth, causing fluid to accumulate in ligated ileal segments in the rabbit and including increased intestinal motility and overt diarrhoea upon oral administration to rhesus monkeys'. The toxic factor is antigenic, susceptible to trypsin and pronase and is thermolabile, being destroyed in 30 min at $56°$. About 80% of the *B. cereus* strains tested by Goepfert and his colleagues have shown evidence of producing this enterotoxic substance. Among other *Bacillus* species tested only *B. thuringiensis* elicited similar pathological symptoms in laboratory animals. Goepfert (1973) considered it was unlikely that the enterotoxic substance is related to the parasporal crystal inclusion body produced by *B. thuringiensis.*

6. Discussion

One can only speculate how many outbreaks of food poisoning would be removed from the 'probable' or 'causal agent unknown' classification if investigators did not just examine foods and clinical specimens for salmonellae, staphylococci and clostridia. Recognizing that a problem exists is an important step but the reporting of such outbreaks will act also as a stimulus to others. This view is supported by the increasing number of outbreaks of food poisoning associated with cooked rice reported in Great Britain since the first account in 1971 and more recently in Canada and Australia. These outbreaks have provided the stimulus for a re-examination of the epidemiological and bacteriological information from outbreaks of food poisoning that could not be attributed to salmonellae or staphylococci.

Table 9 summarizes hitherto unpublished data from 5 outbreaks in England in which an aerobic spore-bearing bacillus was the suspected aetiological agent. Outbreaks 1–4 are from the early records of the Public Health Laboratory Service and outbreak 5 was brought to our attention by Jarvis (pers. comm.). Outbreak 5 is of particular interest because the short incubation period *c.* 4 h and symptoms of vomiting and some diarrhoea are similar to those reported in more recent outbreaks in this country. Also, the custard appeared bitty or granular and was rejected by most of the nurses. The data in Table 9 are incomplete and when outbreaks 1–4 occurred it was not widely appreciated that *Cl. welchii* was a food poisoning agent. Nevertheless, there is good reason to believe that one or more of these outbreaks was caused by aerobic spore-bearing bacilli and that these organisms were *B. cereus.*

7. Future Work

The work on *B. cereus* is still in its early days. There is always a need for accurate and regular reports of as many outbreaks and sporadic cases of food poisoning as possible. In view of the different incubation periods and symptoms recorded and the wide variety of foods implicated, it would be prudent to consider *B. cereus* when investigating all outbreaks of food poisoning. The bacteriological examination of foods from outbreaks should include general aerobic and anaerobic plate counts.

Areas of research include: (1) Monkey feeding tests and volunteer experiments in man using experimentally inoculated foods and culture filtrates of *B. cereus.* The results might confirm the apparent differences in incubation time

MAFF–8

Table 9

Outbreaks of food poisoning in England in which an aerobic spore-bearing bacillus was the suspected aetiological agent

Outbreak	Year	Place	Number ill	Number at risk	Incubation period (h)	Symptoms	Food implicated	Results of bacteriological examination*
1	1940–41	?	?	?	$2\frac{1}{2}$–12 Mean 4	Vomiting, diarrhoea and abdominal pain	Meat (Unspecified)	Large numbers of aerobic spore-bearing bacilli on direct culture of the food; organisms similar to B. megatherium
2	1941	Works canteen	c 100	250	10–12	Diarrhoea, vomiting, colicky pain and slight cramp	Cottage pie	Large numbers of aerobic spore-bearing bacilli on direct culture of the food
3	1941	Works canteen	10	?	12	Severe diarrhoea, nausea but no vomiting	Canned peas	Large numbers $2.5 \times 10^7/g$ of aerobic spore-bearing bacilli in the food
4	1943–45	Hotel (coach party)	?	?	Several hours	Diarrhoea and vomiting	Cold meat	Large numbers $> 10^6/g$ of aerobic spore-bearing bacilli in the food
5	1958	Hospital (nursing staff canteen)	c 50	?	c 4	Vomiting and some diarrhoea	Custard. The custard appeared bitty/granular and was rejected by most of the nurses	Large numbers of aerobic spore-bearing bacilli on direct culture of the food; organisms gave a strong lecithinase reaction

* Cell-free filtrates of the aerobic spore-bearing bacilli from outbreaks 1, 2 and 3 were injected intraperitoneally into kittens. The response was vomiting (outbreak 1), diarrhoea (outbreak 2) and severe vomiting and death (outbreak 3).

and clinical symptoms between outbreaks in different countries. (2) The purification of the enterotoxin described by Goepfert (1973) and a study of its properties. (3) Development of the serological methods briefly referred to in this paper. It may be possible to classify isolates of *B. cereus* as food poisoning strains by using serological or biochemical typing rather than monkey feeding or isolated rabbit ileal loop tests. (4) A study of the survival and growth of *B. cereus* in a variety of foods under various conditions such as temperature, pH and the presence/absence of other micro-organisms.

8. Acknowledgements

We are most grateful to Sir Graham Wilson and Mr. J. D. Jarvis for part of the information included in Table 9 and to Sir Graham who suggested that we should examine the early records of the Public Health Laboratory Service.

We are also grateful to the many Public Health Laboratory Service and Hospital Laboratories in Great Britain, to Dr. J. Taplin, University of Melbourne, Australia and to Dr. J. M. Goepfert, Food Research Institute, University of Wisconsin, U.S.A. for the cultures of *B. cereus* and epidemiological data, and to Dr. Betty C. Hobbs for her advice and encouragement.

9. References

ANON. (1972). Food poisoning associated with *Bacillus cereus. Brit. med J.* **1**, 189.

ANON. (1973). *Bacillus cereus* food poisoning. *Brit. med J.* **3**, 647.

BIRZU, A., GUSITA, C., GALATEANU, M., ONCIU, C. & DINGA, V. (1968). Toxi-infection alimentaire par *B. cereus.* Etude de laboratoire. *Evol. Med.* **12**, 591. Cited in *Biol. Abstr.* 1970, **51**, abstract 91572.

BREKENFELD, H. (1926). Lebensmittelbakterien und Vergiftungen. *Zentbl. Bakt. ParasitKde,* I. Abt. Orig. **99**, 353.

BREKENFELD, H. (1929). Eine Presskopfvergiftungen und ihre Lehren. *Zentbl. Bakt. ParasitKde,* I. Abt. Orig. **110**, 139.

BULYBA, M. S., KUL'CHSKAYA, I. I. & DOMANSKAYA, E. D. (1973). *Bacillus cereus* food poisoning. *Vop. Pitan.* **32**, 86. Cited in *Fd Sci. Technol. Abstr.* 1973, **5**, abstract 6 P 804.

BURDON, K. L. (1956). Useful criteria for the identification of *Bacillus anthracis* and related species. *J. Bact.* **71**, 25.

CHASTAIN, C. B. & HARRIS, D. L. (1974). Association of *Bacillus cereus* with food poisoning in dogs. *J. Am. vet. med. Ass.* **164**, 489.

CHRISTIANSEN, O., KOCH, S. O. & MADELUNG, P. (1951). Et udbrud af levnedsmiddelforgiftning forarsaget af *Bacillus cereus. Nord. VetMed.* **3**, 194.

CLARENBURG, A. & KAMPELMACHER, E. H. (1957). *Bacillus cereus* als oorzaak van voedselvergiftiging. *Voeding,* **18**, 384.

COWAN, S. T. & STEEL, K. J. (1965). *Identification of Medical Bacteria.* Cambridge: Cambridge University Press.

DACK, G. M., SUGIYAMA, H., OWENS, F. J. & KISNER, J. B. (1954). Failure to produce illness in human volunteers fed *Bacillus cereus* and *Clostridium perfringens. J. infect. Dis.* **94**, 34.

DAVIES, F. L. & WILKINSON, G. (1973). *Bacillus cereus* in milk and dairy products. In *The Microbiological Safety of Food.* Eds B. C. Hobbs & J. H. B. Christian. London: Academic Press.

DONOVAN, K. O. (1958). A selective medium for *Bacillus cereus* in milk. *J. appl. Bact.* **21**, 100.

DUNCAN, C. L. & STRONG, D. H. (1971). *Clostridium perfringens* type A food poisoning. I. Response of the rabbit ileum as an indication of enteropathogenicity of strains of *Clostridium perfringens* in monkeys. *Infect. Immunol.* **3**, 167.

FRANKLAND, G. C. & FRANKLAND, P. F. (1887). Studies on some new microorganisms obtained from air. *Phil. Trans. R. Soc. Ser. B.* **178**, 257.

GILBERT, R. J., STRINGER, M. F. & PEACE, T. C. (1974). The survival and growth of *Bacillus cereus* in boiled and fried rice in relation to outbreaks of food poisoning. *J. Hyg., Camb.* **73**, 433.

GOEPFERT, J. M. (1973). Pathogenicity patterns in *Bacillus cereus* food-borne disease. *Abst. 1st Int. Congr. Bacteriol.*, Int. Assoc. Microbiol. Soc., Jerusalem. **1**, 140.

GOEPFERT, J. M., SPIRA, W. M., GLATZ, B. A. & KIM, H. U. (1973). Pathogenicity of *Bacillus cereus*. In *The Microbiological Safety of Food.* Eds B. C. Hobbs & J. H. B. Christian. London: Academic Press.

GOEPFERT, J. M., SPIRA, W. M. & KIM, H. U. (1972). *Bacillus cereus:* food poisoning organism. A review. *J. Milk Fd Technol.* **35**, 213.

HAUGE, S. (1950). Matforgiftninger fremkalt av *Bacillus cereus. Nord. hyg. Tidskr.* **31**, 189.

HAUGE, S. (1955). Food poisoning caused by aerobic spore-forming bacilli. *J. appl. Bact.* **18**, 591.

HAUSCHILD, A. H. W., NIILO, L. & DORWARD, W. J. (1971). The role of enterotoxin in *Clostridium perfringens* type A enteritis. *Can. J. Microbiol.* **17**, 987.

HEINERTZ, N. O. (1962). En nosocomial *B. cereus* intoxication av litet ovanlig karaktär. The Medical Microbiology Division of the Swedish Medical Society, Göteborg. Cited in Nygren (1962).

INAL, T. (1971). Vergleichende Untersuchungen über die Selektivmedien zum qualitativen und quantitativen Nachweis von *Bacillus cereus* in Lebensmitteln. I. *Fleischwirtschaft,* **51**, 1629.

INAL, T. (1972). Vergleichende Untersuchungen über die Selektivmedien zum qualitativen und quantitativen Nachweis von *Bacillus cereus* in Lebensmitteln. II, III and IV. *Fleischwirtschaft,* **52**, 347, 1021, 1160.

JOHANNESEN, S. (1957). Aerobe sporedannende bakterien fra kolonialvarer. *Nord. hyg. Tidskr.* **38**, 231.

KIM, H. U. & GOEPFERT, J. M. (1971*a*). Occurrence of *Bacillus cereus* in selected dry food products. *J. Milk Fd Technol.* **34**, 12.

KIM, H. U. & GOEPFERT, J. M. (1971*b*). Enumeration and identification of *Bacillus cereus* in foods. I. 24 hour presumptive test medium. *Appl. Microbiol.* **22**, 581.

KISS, P. (1961). *Bacillus cereus* altal okozott etelmergezes. *Népegészségügy.* **42**, 87. Cited in *Zentbl. Bakt. ParasitKde,* I. Abt. Ref. 1963, **187**, 217.

LEFEBVRE, A., GREGOIRE, C. A., BRABANT, W. & TODD, E. (1973). Suspected *Bacillus cereus* food poisoning. *Epidem. Bull. Ottawa* **17**, 108.

LUBENAU, C. (1906). *Bacillus peptonificans* als Erreger einer Gastroenteritis-Epidemie. *Zentbl. Bakt. ParasitKde,* I. Abt. Orig. **40**, 433.

MIDURA, T., GERBER, M., WOOD, R. & LEONARD, A. R. (1970). Outbreak of food poisoning caused by *Bacillus cereus. Publ. Hlth. Rep.* **85**, 45.

MORTIMER, P. R. & McCANN, G. (1974). Food poisoning episodes associated with *Bacillus cereus* in fried rice. *Lancet,* i, 1043.

MOSSEL, D. A. A., KOOPMAN, M. J. & JONGERIUS, E. (1967). Enumeration of *Bacillus cereus* in foods. *Appl. Microbiol.* **15**, 650.

NIKODEMUCZ, I. (1958). *Bacillus cereus* als Ursache von Lebensmittelvergiftungen. *Z. Hyg. Infekt-Krankh.* **145**, 335.

NIKODEMUCZ, I. (1965). Die Reproduzierbarkeit der von *Bacillus cereus* verursachten Lebensmittelvergiftungen bei Katzen. *Zentbl. Bakt. ParasitKde,* I. Abt. Orig. **196**, 81.

NIKODEMUCZ, I. (1966*a*). Comparison du pouvoir pathogène expérimental de *Bacillus cereus* et de *Bacillus laterosporus* administrés par voie orale. *Annls Inst. Pasteur, Lille* **17**, 229.

NIKODEMUCZ, I. (1966*b*). Die Wirkung langfristigen Verabreichung von *Bacillus cereus* verunreinigten Lebensmitteln bei Katzen. *Zentbl. Bakt. ParasitKde,* I. Abt. Orig. **199,** 64.

NIKODEMUCZ, I. (1967). Die enteropathogene Wirkung von *Bacillus cereus* bei Hunden. *Zentbl. Bakt. ParasitKde,* I. Abt. Orig. **202,** 533.

NIKODEMUCZ, I. (1968). Die Ätiologie der Lebensmittelvergiftungen in Ungarn in den Jahren 1960 bis 1966. *Z. Hyg. InfektKrankh.* **155,** 204.

NIKODEMUCZ, I., BODNAR, S., BOJAN, M., KISS, M., KISS, P., LACZKO, M., MOLNAR, E. & PAPAY, D. (1962). Aerobe Sporenbildner als Lebensmittelvergifter. *Zentbl. Bakt. ParasitKde,* I. Abt. Orig. **184,** 462.

NIKODEMUCZ, I., MRDODY, Z., SZENTMIHALYI, A. & DOMBAY, M. (1969). Experimente über die enteropathogene Wirkung von *Bacillus. Zentbl. Bakt. ParasitKde.* I. Abt. Orig. **211,** 274.

NIKODEMUCZ, I., NOVOTNY, T., BOUQUET, D. & TARJAN, R. (1967). Über die Ätiologie der Lebensmittelvergiftungen in Ungarn. *Zentbl. Bakt. ParasitKde,* I. Abt. Orig. **203,** 137.

NORRIS, J. R. & WOLF, J. (1961). A study of the antigens of the aerobic spore-forming bacteria. *J. appl. Bact.* **24,** 42.

NYGREN, B. (1962). Phospholipase C-producing bacteria and food poisoning. *Acta path. microbiol. scand. suppl.* **160,** 1.

ORMAY, L. & NOVOTNY, T. (1969). The significance of *Bacillus cereus* food poisoning in Hungary. In *The Microbiology of Dried Foods.* Eds E. H. Kampelmacher, M. Ingram & D. A. A. Mossel. International Association of Microbiological Societies.

ORMAY, L. & NOVOTNY, T. (1970). Über sogenannte unspezifischen Lebensmittel-vergiftungen in Ungarn. *Zentbl. Bakt. ParasitKde,* I. Abt. Orig. **215,** 84.

PISU, I. & STAZZI, L. (1952). Intossicazione alimentare del *Bacillus cereus. Nouvi Annali Ig. Microbiol.* **1,** 1. Cited in Nikodemucz *et al.* (1962).

PIVOVAROV, Yu. P., SIDORENKO, G. I., TKACHENKO, A. V., GOL'DBERG, E. S., AKIMOV, A. M., VOLKOVA, R. S. & SHELAKOVA, V. V. (1970). *Bacillus cereus* as causative organism of food poisoning in man. *Vop. Pitan.* **29,** 25. Cited in *Fd Sci. Technol. Abst.* 1970, **2,** abstract 10C224.

PLAZIKOWSKI, U. (1949). Further investigations regarding the cause of food poisoning. *Rept. Proc. 4th Int. Congr. Microbiol.* 510. Copenhagen: Rosenkilde & Bagger.

SEITZ, M. (1913). Pathogener *Bacillus subtilis. Zentbl. Bakt. ParasitKde,* I. Abt. Orig. **70,** 113.

SMITH, N. R., GORDON, R. E. & CLARK, F. E. (1952). Aerobic sporeforming bacteria. *Agriculture Monograph* No. 16. United States Department of Agriculture.

STONE, M. J. (1952a). The effect of temperature on the development of broken cream. *J. Dairy Res.* **19,** 302.

STONE, M. J. (1952b). The action of the lecithinase of *Bacillus cereus* on the globule membrane of milk fat. *J. Dairy Res.* **19,** 311.

STONE, M. J. & ROWLANDS, A. (1952). 'Broken' or 'bitty' cream in raw and pasteurized milk. *J. Dairy Res.* **19,** 51.

TAYLOR, A. J. & GILBERT, R. J. (1975). *Bacillus cereus* food poisoning: a provisional serotyping scheme. *J. med. Microbiol.* **8,** 543.

TODD, E., PARK, C., CLECNER, B., FABRICUS, A., EDWARDS, D. & EWAN, P. (1974). Two outbreaks of *Bacillus cereus* food poisoning in Canada. *Can. J. publ. Hlth.* **65,** 109.

TRÜB, C. L. P. & WUNDRAM, G. (1942). Die Gemeinschaftsverpflegung in ihrer Beziehung zu den unspezifischen bakteriellen Lebensmittelvergiftungen. Volksgesundh. Reichminist. Innern 56 No. 1. Cited in *Zentbl. Bakt. ParasitKde,* I. Abt. Ref. 1943, **143,** 241.

VLAD, A. & VLAD, A. (1972). Toxiinfectie alimentară provocată de *Bacillus cereus. Microbiologia Parazit. Epidem.* **17,** 531.

VON LUDWIG, K. (1971). *Bacillus cereus* als Ursache einer Lebensmittelintoxikation. *Arch. Lebensmittelhyg.* **22,** 104.

Trends in Methods for Detecting Food-poisoning Toxins Produced by *Clostridium botulinum* and *Staphylococcus aureus*

J. S. CROWTHER AND R. HOLBROOK

*Unilever Research, Colworth House,
Sharnbrook, Bedford, England*

CONTENTS

1. Introduction

STAPHYLOCOCCAL ENTEROTOXINS are toxic proteins produced by a large proportion of strains of *Staphylococcus aureus* and are the causative agents of staphylococcal food poisoning. To date 5 serologically distinct enterotoxins have been identified, designated A–E. Some strains of *Staph. aureus* produce monkey emetic substances that cannot be identified by using antisera to any of these 5 serological types (Bergdoll, 1972). Types A and D are most frequently implicated in food poisoning incidents (Casman *et al.*, 1967; Toshach & Thorsteinson, 1972; Gilbert & Wieneke, 1973). The minimum dose of enterotoxin that can cause food poisoning in man is not known, but Bergdoll (1973) and Gilbert & Wieneke (1973) consider that as little as 1 μg may be sufficient. If this quantity of toxin is consumed in 100 g of food, then toxin concentrations in the order of 0.01 μg/g of food are hazardous. Therefore, methods which will detect enterotoxin concentrations in the order of 0.001 μg/g seem necessary. This concentration cannot yet be detected with confidence in all types of food.

Clostridium botulinum is known to produce 7 serologically distinct neurotoxins, designated A–G. Types A, B, E and F can all cause fatal food poisoning in man; type C intoxicates birds and type D, cattle. Type G, recently isolated from soil by Gimenez & Ciccarelli (1970) has not been implicated in disease of

215

man or animals. Again the toxic dose for man is not known, but Boroff & DasGupta (1971) suspect that as little as 1 μg may be fatal. The botulinum toxins have been reviewed thoroughly by Boroff & DasGupta (1971) and Schantz & Sugiyama (1974).

Although the food-poisoning toxins of *Staph. aureus* and *Cl. botulinum* are very different in nature and mode of action, they pose similar problems to the diagnostic bacteriologist trying to detect them. Toxins of both organisms are water-soluble proteins which can be present in very small amounts in food containing very large amounts of other proteins. Thus, either sensitive techniques or efficient extraction and concentration procedures are needed to detect them. Both toxins are antigenic and can be detected serologically.

The most useful attributes for an ideal test system for either botulinum toxin or staphylococcal enterotoxin seem to be: (1) specific reactions, no interference or cross-reactions from other organisms or food constituents; (2) sensitive, to detect toxin at concentrations lower than those associated with intoxication; (3) rapid, results within one working day; (4) polyvalent antisera required to cover all serological types; (5) reagents should be stable and inexpensive; (6) animal tests should be avoided; (7) procedures could be automated.

In this paper we review recent attempts by ourselves and others to improve the speed, sensitivity and specificity of *in vitro* methods for detecting these toxins.

2. Accepted Extraction and Detection Methods

The accepted method for detecting botulinum toxin is the mouse test, in which a sample of food is extracted with buffer containing gelatin, and injected intraperitoneally into mice without further concentration (Baird-Parker, 1969). Mice protected with specific antitoxins are used as controls. The mouse test is simple, but it has the disadvantage that non-specific deaths may occur due to protein shock and other causes. There is, then, the need for a more rapid and specific test to detect botulinum toxins.

At present the most widely used method for detecting staphylococcal enterotoxins is the extraction and concentration procedure of Casman (1967), followed by serological detection of the enterotoxin using the Crowle microslide technique (Casman, Bennett, Dorsey & Stone, 1969). This extraction method is cumbersome and time-consuming, because in order to achieve the required sensitivity, the extract from 100 g of food has to be concentrated to *c.* 0.2 ml; a minimum of 3–4 days is needed for extraction, and a further 1–3 days for serological detection (Table 1). Modifications to this method have been made by different workers either to reduce the processing time or to adapt the method for use with a particular type of food (Zehren & Zehren, 1968; Gilbert *et al.*, 1972). Recently a more rapid method has been described by Reiser, Conaway &

Table 1

Examination of foods for staphylococcal enterotoxins

Time	Procedure
1st day	Extract toxin from food with 0.2 M NaCl, pH 7.4
	Remove lipids with chloroform
Overnight	Dialyse against polyethylene glycol to concentrate toxin and remove salts
2nd and 3rd days	Separate toxin from other proteins by carboxymethylcellulose chromatography
Overnight	Concentrate to < 0.5 ml by dialysis
4th day	Re-extract with chloroform, centrifuge
	Examine serologically for enterotoxin
5th–7th day	*RESULT*

After Casman & Bennett (1965).

Bergdoll (1974), which can be completed in 2 days, but still using ion-exchange chromatography.

Several precipitin techniques have been described for detecting staphylococcal enterotoxins in the extracts, including both single and double diffusion methods. These have been reviewed by Bergdoll (1970). Of these the Crowle micro double-diffusion slide technique is probably most widely established, and Holbrook & Baird-Parker (1975) have compared it in detail with other available techniques. The advantage of the Crowle method is that only small quantities (0.02 ml) of both antisera and test material are required thereby conserving antiserum which is difficult to prepare. A further advantage is that proven identity of the test toxin can be confirmed by the use of control toxin in an adjacent well. Its disadvantage is the incubation period necessary for precipitin lines to develop from low levels of toxin — at least one day at 37° or 3 days at 20°.

3. Advances in Methods for Detecting Toxins

(a) *Precipitin techniques*

(i) *Counter-current immuno-electrophoresis*

Serological precipitin techniques using passive diffusion for detecting botulinal toxins have not been successful, because of their lack of sensitivity. Vermilyea, Walker & Ayres (1968) using a slide modification of the Ouchterlony double-diffusion method, could not detect < 370 MLD of type B toxin/ml of test material. However, in recent years a modification of the Ouchterlony double-diffusion method has been developed which is claimed to detect rapidly very small amounts of bacterial antigen. This is counter-current electro-immunodiffusion or 'cross-over' electrophoresis. The principle of the technique is identical to that of the conventional Ouchterlony method except that the antigen and antibody are made to migrate actively towards each other in an electric field. The great advantage of the method is that precipitin lines appear in only 30–60 min compared with 1–3 days for the conventional method. A further advantage is that as many as 15 samples can be tested on one microscope slide. The method has been used successfully in clinical laboratories to detect and type antigens of meningococci (Edwards, 1971), pneumococci (Dorff, Coonrod & Rytel, 1971) and other streptococci (Edwards & Larson, 1973), in patients' serum or cerebro-spinal fluids. Cross-over electrophoresis has been developed by Kimble & Anderson (1973) to detect staphylococcal enterotoxin A. This was accomplished after introducing sulphonated phenyl groups into the immune γ-globulin to increase its net negative charge, without loss of serological activity. The method produced precipitin lines in 1–2 h but the sensitivity was only 3–8 μg of enterotoxin A/ml.

We have examined counter-current electrophoresis to detect botulinum type A toxin in culture supernatant fractions (Crowther & Baird-Parker, unpublished). Antitoxin was raised in rabbits against crystalline botulinum type A toxin and used in 'cross-over' electrophoresis gels to detect toxins. When crystalline toxin (i.e. homologous toxin) was used as the antigen, as little as 40 MLD/ml could be detected. However, when culture supernatant liquids of 50 strains of *Cl. botulinum* from a wide variety of sources were tested, only 13 of 50 gave precipitin lines, even though all were shown to be toxic for mice. Some cultures still did not give lines after concentrating them 100-fold by vacuum dialysis. Thus in our hands, cross-over electrophoresis was not reliable for detecting botulinum type A toxin.

(ii) *Electro-immunodiffusion*

Another electrophoresis-gel precipitin test which has been used to detect both staphylococcal (Chugg, 1972; Gasper, Heimsch & Anderson, 1973) and botulinum (Miller & Anderson, 1971) toxins is electro-immunodiffusion or Laurell electrophoresis (Laurell, 1966). In this technique the antitoxin is incorporated into the agarose gel. The toxin is introduced into a well cut in the agarose, and under the influence of an electric field, it migrates from the well through the agarose. A precipitate in the shape of a cone is formed where toxin and antitoxin concentration are equivalent. The height of the precipitin cone is proportional to the concentration of antigen and inversely proportional to the antibody concentration.

Although electro-immunodiffusion is a rapid method it is not as sensitive as the mouse test for botulinum toxins. Miller & Anderson (1971) could detect down to only 140 MLD of type A toxin/ml. We could detect down to only

512 MLD of type E toxin/ml in foods contaminated deliberately with *Cl. botulinum* type E (Crowther & Baird-Parker, unpublished).

Electro-immunodiffusion has also been used as a rapid method for the detection of staphyloccocal enterotoxins, and we have established conditions which permitted the detection of 0.5 μg of toxin/ml (Holbrook & Baird-Parker, 1975). At this concentration of toxin the precipitate required staining for its detection. Gasper, Heimsch & Anderson (1973) increased the sensitivity to 0.15 μg/ml by increasing the sample volume and by developing the precipitin complex by reaction with sheep anti-rabbit IgG, precipitation with cadmium acetate and staining. A disadvantage of the method is that other food proteins present in the test sample will mask weak reactions unless these are removed by extensive washing in salt solution.

(b) *Reverse passive haemagglutination*

In the reverse passive haemagglutination technique, purified antibody (γ-globulin) is coupled to erythrocytes using tannic acid or other coupling agents. Test material is added and, if toxin is present, the sensitized erythrocytes are agglutinated. The method has been used to detect enterotoxin A and B and botulinum type A, B and E toxins. The method is very sensitive, detecting 0.001 μg of enterotoxin/ml and gives results within 4 h (Silverman, Knott & Howard, 1968). However, high-titred specific antiserum is required and a rigorously standardized technique is necessary to obtain consistent results. Unfortunately, the method has not proved entirely satisfactory for the detection of enterotoxin because some food proteins cause false positive results (Bennett *et al.*, 1973). Bergdoll and co-workers have partially overcome this by treating the food extracts with trypsin to digest the interfering proteins; this is possible because enterotoxins are generally resistant to proteolytic enzymes (Bergdoll, 1973). Both Bergdoll and ourselves have found this method difficult to use to detect enterotoxin A, because γ-globulin from some rabbits appears to be either inactivated by the coupling procedure or not adsorbed to the red blood cells.

Reverse passive haemagglutination has been examined by several workers as a test for botulinum toxins. Sinitsyn (1960) found the technique more sensitive than the mouse test for *Cl. botulinum* types A and B toxin, but false negative results occurred frequently. Johnson *et al.* (1966) improved the technique considerably and could detect < 3 mouse LD_{50} of A or B toxin in culture supernatant fractions or food samples without false negative results; Evancho *et al.* (1973) detected 27 MLD/ml of type A toxin without cross-reactions from other types.

All these experiments have been made using antitoxin prepared from crystalline botulinum toxin. This approach, however, has recently come under strong criticism from Sugiyama, Ohishi & DasGupta (1974) on the grounds that the reverse passive haemagglutination tests using antitoxin prepared against crystalline botulinum toxin do not test specifically for *toxin*, but may instead measure non-toxic haemagglutinin with which the toxin is intimately associated. This criticism can be best understood by considering what is known about crystalline botulinum type A toxin. Botulinum type A toxin consists of 2 distinct proteins, a toxin and a haemagglutinin. Under most conditions the 2 proteins form a stable complex which appears chemically homogenous, but the 2 proteins can be separated by ion-exchange chromatography to give the toxin and haemagglutinin. The relative proportions of toxin and haemagglutinin moieties in any preparation can vary considerably, but in all analyses reported so far the haemagglutinin has been found to be the dominant component (Boroff & DasGupta, 1971). Sugiyama, Ohishi & DasGupta (1974) have shown that antitoxin prepared from crystalline toxin has a higher titre to the haemagglutinin component than to the toxin. Thus, when red blood cells were sensitized with this antitoxin, the antibodies coupled to them were shown to be almost all haemagglutinin antibodies, and the reverse passive haemagglutination test would therefore assay botulinum haemagglutinin and not toxin. Most of the detection methods described in the literature would therefore be invalidated as reliable methods for detecting toxin. Sugiyama and his colleagues emphasize that any serological test designed to detect botulinum toxin must be one in which the antitoxin is prepared from toxin freed from haemagglutinins. So far only types A, B, D and E toxins have been so purified. This criticism seems logical and must point the way for future serological tests for botulinum toxins.

(c) *Radio-immunoassay*

Solid phase radio-immunoassay has been developed by 2 independent groups of workers in the U.S.A. for detecting enterotoxins A and B. In both methods a known amount of purified enterotoxin IgG is attached to a solid carrier, such as polystyrene tubes (Johnson *et al.*, 1971) or bromacetylcellulose (Collins, Metzger & Johnson, 1972). Simple food extracts are mixed with a known amount of ^{125}I-labelled toxin and added to the immobilized antibody. The labelled toxin and toxin in the food extract compete for binding sites on the antibody. The levels of bound or free ^{125}I-labelled antigen are detected using γ-particle counting equipment and the level of toxin in the food extract is calculated from a standard curve. This method has now been shown to detect enterotoxins A and B at concentrations between 0.01 and 0.001 μg/g in several foods; these are encouraging results. Its disadvantages are those limitations imposed by the use of radioactive materials, the need for expensive counting equipment, and highly purified enterotoxin labelled with ^{125}I. The method also requires highly purified specific antiserum.

A radio-immunoassay has been developed by Boroff & Shu-Chen (1973*a,b*) to detect botulinum type A toxin. Their procedure uses antitoxin raised from haemagglutinin-free toxin so undeniably detects toxin. Boroff & Shu-Chen claim that the method can detect down to 5–10 MLD of type A toxin. This success has prompted these workers to develop the test for other toxigenic types of *Cl. botulinum.*

4. Conclusions

Sensitive methods for detecting antigen or antibody systems which have been developed in other areas of immunology, have been applied to the detection of staphylococcal enterotoxin and botulinum toxin.

Table 2 summarizes the various methods for detecting staphylococcal enterotoxin and shows how they meet the requirements of the ideal test. The Crowle micro-slide method can be used to detect all 5 serological types, because antisera with sufficient specificity have been prepared for each type. The method however requires lengthy incubation. The electrophoretic methods which have been devised to overcome this problem lack the required sensitivity. However these serological precipitin methods also lack required sensitivity because of our inability to detect precipitates at toxin concentrations below c. 0.1 μg/ml. At the present time the reverse passive haemagglutination technique meets most of the ideal requirements but unless the problems of non-specific reactions with some food constituents and conjugation of antibody to erythrocytes is overcome the method cannot be reliable. Radio-immunoassay also meets many of the criteria but has the disadvantage of using radioactive material and requires highly purified enterotoxins which at present have not been prepared from all 5 serological types.

None of the precipitin tests in agar gel show sufficient sensitivity to replace mouse toxicity as a routine test for botulinum toxins (Table 3). Reverse passive haemagglutination techniques are relatively simple and offer good sensitivity, but they have not been perfected for all toxigenic types. Radio-immunoassay promises to have the required sensitivity but has the disadvantages of requiring pure antitoxin and labelled pure toxin. Nevertheless, it could become a routine method for the future.

5. Summary

This paper reviews recent attempts by ourselves and others to improve the speed, sensitivity and specificity of *in vitro* methods for detecting botulinum and staphylococcal food-poisoning toxins. Precipitin methods including double gel-diffusion, electro-immunodiffusion and counter-current immuno-electrophoresis ('cross-over' electrophoresis) lack sufficient sensitivity to detect the presumed human toxic doses, without preliminary concentration of extracts.

Table 2

Comparison of methods for enterotoxin detection

Method	Limit of detection µg/ml	Used to detect toxins	Approx. time for result	Disadvantage
Crowle micro double-diffusion	0.2–0.1	A to E	2–3 days	Slow, lacks sensitivity
Electro-immunodiffusion	{ 0.15 0.5	A A & B	> 1 day < 1 day	Lacks sensitivity
Counter-current immuno-electrophoresis	1.0	A	1 day	Modification of globulin to increase negative charge
Reverse passive haemagglutination	0.001	A & B	< 1 day	Requires high titred specific antiserum
Radio-immunoassay	0.001	A & B	< 1 day	Needs radio-labelled pure toxin

Data from: Silverman, Knott & Howard (1968); Johnson et al. (1971); Collins, Metzger & Johnson (1972); Bergdoll (1973); Gasper, Heimsch & Anderson (1973); Kimble & Anderson (1973); Holbrook & Baird-Parker (1975).

Table 3

Comparison of methods for detection of botulinum toxins

Method	Approx. limit of detection (MLD/ml)	Used to detect toxins	Approx. time for result	Disadvantage
Mouse toxicity	1	A to G	1–3 days	Some false positives; may be slow
Double diffusion	370	B	1–2 days	Concentration of sample needed. False negatives
Counter-current immuno-electrophoresis	100	A	3–4 h	Some false negatives
Electro-immunodiffusion	140	A	2–3 h	
Reverse passive haemagglutination	{27 / 10	A / E	1–2 h	Some cross-reactions between E and A
Radio-immunoassay	5–10	A	not quoted	Needs radio-labelled pure toxin

Data from: Vermilyea, Walker & Ayres (1968); Miller & Anderson (1971); Boroff & Shu-Chen (1973*a,b*); Evancho *et al.* (1973); Crowther & Baird-Parker (unpublished).

Reverse passive haemagglutination and radio-immunoassay methods are more sensitive and show promise as rapid methods for the future.

6. Acknowledgements

We gratefully acknowledge Dr. P. D. Walker, Wellcome Research Laboratories, Beckenham for his kind help and advice on how to prepare botulinum toxins and anti-toxins, and Professor M. S. Bergdoll, Food Research Institute, Wisconsin, for supplying reagents to initiate our own studies on staphylococcal enterotoxins.

7. References

BAIRD-PARKER, A. C. (1969). Medical and veterinary significance of spore-forming bacteria. In *The Bacterial Spore*. Eds G. W. Gould & A. Hurst. New York: Academic Press.

BENNETT, R. W., KEOSEYAN, S. A., TATINI, S. R., THOTA, H. & COLLINS, W. S. (1973). Staphylococcal enterotoxin: a comparative study of serological detection methods. *J. Inst. Can. Sci. Technol.* **6**, 131.

BERGDOLL, M. S. (1970). Enterotoxins. In *Microbial Toxins* Vol. III. Eds T. C. Montie, S. Kadis & S. T. Ajl. New York: Academic Press.

BERGDOLL, M. S. (1972). The Enterotoxins. In *The Staphylococci*. Ed. J. O. Cohen. New York: Wiley Interscience.

BERGDOLL, M. S. (1973). Enterotoxin detection. In *The Microbiological Safety of Food*. Eds B. C. Hobbs & J. H. B. Christian. New York: Academic Press.

BOROFF, D. A. & DASGUPTA, B. R. (1971). Botulinum toxin. In *Microbial Toxins,* Vol. IIA. Eds S. Kadis, T. C. Montie & S. J. Ajl. New York: Academic Press.

BOROFF, D. A. & SHU-CHEN, G. (1973a). Radioimmunoassay for type A toxin of *Clostridium botulinum. Appl. Microbiol.* **25**, 545.

BOROFF, D. A. & SHU-CHEN, G. (1973b). Radioimmunoassay of toxin of *Clostridium botulinum. Fed. Proc.* **32**, 1032.

CASMAN, E. P. (1967). Staphylococcal food poisoning. *Hlth. Lab. Sci.* **4**, 199.

CASMAN, E. P. & BENNETT, R. W. (1965). Detection of staphylococcal enterotoxin in food. *Appl. Microbiol.* **13**, 181.

CASMAN, E. P., BENNETT, R. W., DORSEY, A. E. & ISSA, J. A. (1967). Identification of a fourth staphylococcal enterotoxin; enterotoxin D. *J. Bact.* **94**, 1875.

CASMAN, E. P., BENNETT, R. W., DORSEY, A. E. & STONE, J. E. (1969). The microslide gel double diffusion test for the detection and assay of staphylococcal enterotoxins. *Health Lab. Sci.* **6**, 185.

CHUGG, L. R. (1972). Rapid sensitive methods for enumeration and enterotoxin assay for *Staphylococcus aureus.* Ph.D. Thesis, Oregon State University *Dissert. Abs.* **32**, 4383 B.

COLLINS, W. S. II, METZGER, J. F. & JOHNSON, A. D. (1972). A solid phase radioimmunoassay for staphylococcal B enterotoxin. *J. Immunol.* **108**, 852.

DORFF, G. J., COONROD, J. D. & RYTEL, M. W. (1971). Detection by immunoelectrophoresis of antigen in sera of patients with pneumococcal bacteraemia. *Lancet* **1**, 578.

EDWARDS, E. A. (1971). Immunologic investigation of meningococcal disease. *J. Immunol.* **106**, 314.

EDWARDS, E. A. & LARSON, G. L. (1973). Serological grouping of hemolytic streptococci by counter-immunoelectrophoresis. *Appl. Microbiol.* **26**, 899.

EVANCHO, G. M., ASHTON, D. H., BRISKEY, E. J. & SCHANTZ, E. J. (1973). A standardized reversed passive haemagglutination technique for the determination of botulinum toxin. *J. Fd. Sci.* **38**, 764.

GASPER, E., HEIMSCH, R. C. & ANDERSON, A. W. (1973). Quantitative detection of type A staphylococcal enterotoxin by Laurell electroimmunodiffusion. *Appl. Microbiol.* **25**, 421.

GILBERT, R. J. & WIENEKE, A. A. (1973). Staphylococcal food poisoning with special reference to the detection of enterotoxin in food. In *The Microbiological Safety of Food.* Eds B. C. Hobbs & J. H. B. Christian. London & New York: Academic Press.

GILBERT, R. J., WIENEKE, A. A., LANSER, J. & ŠIMKOVIČOVÁ, M. (1972). Serological detection of enterotoxin in food implicated in staphylococcal food poisoning. *J. Hyg., Camb.* **70**, 755.

GIMENEZ, D. F. & CICCARELLI, A. S. (1970). Another type of *Clostridium botulinum.* *Zentbl. Bakt. ParasitKde* Abt. I, Orig. **215**, 221.

HOLBROOK, R. & BAIRD-PARKER, A. C. (1975). Serological methods for the assay of staphylococcal enterotoxins. In *Some Methods for Microbiological Assay.* Soc. Appl. Bact. Technical Series No. 8. Eds R. G. Board & D. W. Lovelock. London & New York: Academic Press.

JOHNSON, H. M., BRENNER, K., ANGELOTTI, R. & HALL, H. E. (1966). Serological studies of types A, B and E botulinal toxins by passive haemagglutination and bentonite flocculation. *J. Bact.* **91**, 967.

JOHNSON, H. M., BUKOVIC, J. A., KAUFFMAN, P. E. & PEELER, J. T. (1971). Staphylococcal enterotoxin B; Solid-phase radioimmunoassay. *Appl. Microbiol.* **22**, 837.

KIMBLE, C. E. & ANDERSON, A. W. (1973). Rapid, sensitive assay for staphylococcal enterotoxin A by reversed immuno-osmophoresis. *Appl. Microbiol.* **25**, 693.

LAURELL, C. B. (1966). Quantitative estimation of proteins by electrophoresis in agrose gel containing antibodies. *Ann. Biochem.* **15**, 45.

MILLER, C. & ANDERSON, A. W. (1971). Rapid detection and quantitative estimation of type A botulinum toxin by electroimmunodiffusion. *Infect. Immun.* **4**, 126.

REISER, R., CONAWAY, D. & BERGDOLL, M. S. (1974). Detection of staphylococcal enterotoxins in foods. *Appl. Microbiol.* **27**, 83.

SCHANTZ, E. J. & SUGIYAMA, H. (1974). Toxin proteins produced by *Clostridium botulinum. J. Agric. Fd. Chem.* **22**, 26.

SILVERMAN, S. J., KNOTT, A. R. & HOWARD, M. (1968). Rapid sensitive assay for staphylococcal enterotoxin and a comparison of serological methods. *Appl. Microbiol.* **16**, 1019.

SINITSYN, V. A. (1960). Use of indirect haemagglutination reaction in detection of botulinic toxins. *J. Microbiol. Epid. Immun.* **31**, 408.

SUGIYAMA, H., OHISHI, I. & DASGUPTA, B. R. (1974). Evaluation of type A botulinal toxin assays that use antitoxin to crystalline toxin. *Appl. Microbiol.* **27**, 333.

TOSHACH, S. & THORSTEINSON, S. (1972). Detection of staphylococcal enterotoxins by the gel diffusion test. *Can. J. publ. Hlth.* **63**, 58.

VERMILYEA, B. L., WALKER, H. W. & AYRES, J. C. (1968). Detection of botulinum toxins by immunodiffusion. *Appl. Microbiol.* **16**, 21.

ZEHREN, V. L. & ZEHREN, V. F. (1968). Examination of large quantities of cheese for staphylococcal enterotoxin A. *J. Dairy Sci.* **51**, 635.

The Exploitation of Transmissible Plasmids to Study the Pathogenesis of *Escherichia coli* Diarrhoea

H. Williams Smith

Houghton Poultry Research Station, Houghton,
Huntingdon PE17 2DA, England

CONTENTS

1. Introduction

THE IMPORTANCE of transmissible plasmids has only been generally recognized during the last 15 years in which they have been shown to be responsible for most antibiotic resistance existing in aerobic Gram negative bacteria that inhabit the alimentary tract of man and domestic animals. Until that time, all the resistance exhibited by these bacteria was thought to be mutational in origin. Unlike the genes coding for mutational resistance, which are located on the chromosome, those responsible for transmissible resistance are located extrachromosomally as plasmids. Moreover, they can be transmitted from one bacterial cell to another by conjugation, a process that is also plasmid-controlled. Because plasmids are autonomous, cells that acquire them during conjugation may subsequently lose them, a process that can sometimes be accelerated in the laboratory by treatment with agents such as acridine orange, ethidium bromide and sodium lauryl sulphate. Thus by laboratory manipulations, employing conjugation procedures and treatment with these agents, it is possible, within limits, to create lines of bacteria of a desired plasmid composition.

In recent years, characteristics of *E. coli* other than antibiotic resistance have been found to be controlled by transmissible plasmids. This is especially so in the case of those strains that produce diarrhoea in man and domestic animals, the so-called enteropathogenic strains. The purpose of this paper is to show how, by recourse to plasmid manipulation, this fact can be exploited to study the pathogenesis of *E. coli* diarrhoea.

2. The Essential Bacteriological Features of *Escherichia coli* Diarrhoea

Neonatal diarrhoea due to *E. coli* occurs in human babies, piglets, calves and lambs. The disease is essentially similar in all 4 species: in humans and pigs different serotypes of *E. coli* are responsible for the disease whereas those affecting calves and lambs are common to both species. Thus, conclusions reached from studying the disease in one species may be equally applicable to the disease as it occurs in the other species.

The essential bacteriological features of *E. coli* diarrhoea in all species is exemplified by the results, summarized in Table 1, of estimating the concentrations of viable *E. coli* organisms in the contents of different parts of the

Table 1

Escherichia coli *content of the alimentary tract in 8 pairs of piglets, one ill and one apparently normal, from different outbreaks of diarrhoea*

Status of piglets	Log_{10} number of *E. coli* organisms/g* of chyme in					
	Stomach	Small intestine portion				Colon
		1	3	5	7	
Diseased	2.7	7.3	8.5	9.4	9.5	9.6
Healthy	2.2	3.6	4.8	7.3	8.3	9.0

* The median count for 8 piglets is given. The small intestine was divided into seven equal portions; portion 1 was nearest to the stomach and portion 7 nearest to the colon.
(From Smith & Jones, 1963.)

alimentary tract of 8 pairs of piglets from different outbreaks of *E. coli* diarrhoea; each pair consisted of a diarrhoeic piglet and an apparently normal piglet from the same litter. The important difference between the diarrhoeic piglets and the normal piglets was the enormous proliferation of *E. coli* in the small intestine, particularly in the anterior part, of the diarrhoeic piglets. The proliferating organisms in each of these diarrhoeic piglets belonged to serotypes known to be enteropathogenic for pigs (Sojka, 1971). There was no evidence of invasion of the body by any of these organisms; neither was there any obvious disorganization of the general bacterial flora of the alimentary tract in the diseased group suggestive of alimentary dysfunction.

In view of these facts, the first question that arises is how do the pathogenic *E. coli* proliferate in the anterior regions of the small intestine and avoid being flushed through with the chyme? They probably adhere to the epithelium of the small intestine so that the question now becomes: how do they achieve this? The next important question is how do the proliferating organisms cause the enormous outpouring of fluid from the body into the small intestine which shows as severe diarrhoea and dehydration?

3. Transmissible Plasmids in Enteropathogenic *E. coli*

Enteropathogenic strains, are classified serologically according to their O, K and H antigens. Those producing diarrhoea in babies are recorded as possessing one K antigen e.g. O26 : K60 and O55 : K59 and so are a few enteropathogenic for pigs, e.g. O138 : K81 and O141 : K85ac. Most pig enteropathogenic strains, however, possess 2 K antigens, one of which, K88, either in its ab or ac form, being common to all of them, e.g. O141 : K85ab, 88ab and O8 : K87, 88ac. The K88 antigen is controlled by a transmissible plasmid (Orskov & Orskov, 1966). Recently, Sojka (unpublished) reported that most strains enteropathogenic for calves and lambs also possess 2 K antigens, one, which has been given the International designation K99, being common to all of them, e.g. O8 : K85,99 and O101 : K32,99. The K99 antigen, too, is plasmid-controlled (Smith & Linggood, 1971b). Most of the pig enteropathogenic strains produce a filterable haemolysin and this in some strains is controlled by a transmissible plasmid designated Hly (Smith & Halls, 1967a). *Escherichia coli* strains pathogenic for babies, pigs, calves and lambs all produce enterotoxin, this being defined originally as a substance that causes dilatation of ligated segments of small intestine (Plate 1). The enterotoxin produced by some pig strains (Smith & Halls, 1968a), calf and lamb strains (Smith & Linggood, 1972) and human strains (Smith & Linggood, 1971a; Skerman, Formal & Falkow, 1972) have been shown to be plasmid-controlled, the plasmid responsible being designated Ent. The enterotoxins of pig strains exist in 2 forms, a heat-labile form (LT) and a heat-stable form (ST). Some strains produce LT and ST and others produce ST only; both forms are plasmid-controlled (Smith & Gyles, 1970). It is apparent then that the enterotoxin of human, pig, calf and lamb strains can be utilized in plasmid manipulation studies. So can the K99 antigen of calf and lamb strains and the K88 antigen and the haemolysin of pig strains.

4. Technique of Plasmid Manipulation

The method of transmitting the Ent (enterotoxin), Hly (haemolysin), K88 and K99 plasmids consists of growing in broth a donor strain possessing one or more of these plasmids and a recipient strain which is a chromosomal mutant resistant to streptomycin (or to some other antibiotic). This mixed culturing permits conjugation and plasmid transfer to occur. To detect recipient organisms that have acquired Hly, K88 and K99, the mixed culture is then plated on washed sheep blood agar containing streptomycin to suppress donor organisms. After incubation, the colonies are examined visually for haemolysin production (Plate 2) and by slide agglutination tests with appropriate antisera for K88 or K99 production.

For detecting Ent transfer, the mixed cultures are often grown twice in broth containing streptomycin to eliminate donor organisms and then once in plain broth. The purpose of this technique is to provide maximal opportunity for plasmid transfer since tests for Ent activity are usually performed in ligated segments of small intestine of pigs, calves and rabbits there is a severe limit to the number that can be performed. Furthermore, the final culture itself is usually tested in a ligated segment and only if it gives a positive result are cultures of single colonies obtained with the object of finding one that is Ent$^+$. The results of testing mating cultures in pig intestine are illustrated in Plate 3. Those for tests in rabbit intestine of bacteria-free LT preparations from colonies obtained from a positive mating culture are illustrated in Plate 4; LT preparations instead of live cultures were employed here because the recipient was *E. coli* K12, a strain that grows poorly in ligated intestinal segments. The donor in Plate 4 was a human strain; those in Plate 3 were pig strains.

The removal of Hly, K88 and K99 plasmids can be achieved, by growing organisms possessing them in broth containing acridine orange, ethidium bromide or sodium lauryl sulphate (Smith & Linggood, 1971*b*) and then plating them on suitable media and examining colonies for loss of these characters. Ent$^-$ organisms could not be obtained from Ent$^+$ strains by these or other methods.

5. The Oral Administration of LT and ST-type Preparations of Ent$^+$ and Ent$^-$ Forms of the same Strain of *E. coli* to Experimental Animals

The results of giving piglets, orally, bacteria-free LT-type and ST-type preparations of non-pathogenic strains of *E. coli* that had either received the Ent plasmid or not is summarized in Table 2. The LT-preparations of the Ent$^+$ recipient strains produced severe diarrhoea within 3–5 h of administration and the ST-type within 2 h. It lasted for 7–24 h and was accompanied by

Table 2

The results of oral administration of enterotoxin preparations to piglets

Enterotoxin preparation administered		No. of piglets	
Type	*E. coli* source	treated	that developed diarrhoea
LT	Ent$^+$ recipient strain	18	16
(heat-labile)	Ent$^-$ recipient strain	18	2
ST	Ent$^+$ recipient strain	16	16
(heat stable)	Ent$^-$ recipient strain	16	0
Untreated		56	0

Each piglet was given 20–40 ml of enterotoxin preparation.
(Modified from Smith & Gyles, 1970.)

pronounced clinical signs of dehydration; despite treatment several of the piglets died. The diarrhoea produced in the two piglets given LT-type preparations of an Ent$^-$ strain was mild, transient and unaccompanied by constitutional disturbances.

Similarly, baby rabbits given bacteria-free preparations of Ent$^+$ forms of the non-pathogenic *E. coli* K12 strain developed severe diarrhoea, often terminating in death, whereas their litter mates given preparations of Ent$^-$ forms of the same strain remained healthy. The severity of the diarrhoea was similar to that produced by cholera toxin. A human enteropathogenic strain, in addition to a pig enteropathogenic strain, was used as an Ent donor in these experiments. The results of some of these experiments are illustrated in Plates 5 and 6. The baby rabbits, like the piglets, had only been removed from their mothers for the minute it took to inoculate them by stomach tube.

Because the 2 kinds of materials given to the piglets and baby rabbits were prepared from organisms that were isogenic apart from the fact that they did or did not possess an Ent plasmid, these experiments provide strong evidence that enterotoxin is important in the pathogenesis of *E. coli* diarrhoea.

6. The Oral Administration to Animals of Live Cultures of Organisms Subjected to Plasmid Manipulation

In these experiments, naturally-reared animals were used. The pigs belonged to a herd specially in-bred because they were susceptible to *E. coli* infection. Piglets, all from primiparous sows, were infected in their first day of life after they had received colostrum; pigs were infected a few days after weaning when 8 weeks old. Unless stated, infection was achieved by the oral administration of broth or nutrient agar cultures (for fuller details see Smith & Linggood, 1971*b*).

(a) *The significance of the Hly and K88 plasmids in the*
ability of an enteropathogenic O141 : K85ab, 88ab strain of
E. coli *to infect pigs*

The results of giving piglets a wild, pig enteropathogenic strain possessing both the Hly and K88 plasmids, or forms of it from which one or other of these plasmids had been eliminated, are summarized in Table 3. Likewise the results of giving the Hly$^+$ K88$^-$ form after the introduction of another K88ab plasmid, this new K88 being derived from an O8 : K87, 88ab strain. It was transmitted from this strain to an *E. coli* K12 strain and from the latter to the Hly$^+$ K88$^-$ form of the O141 strain. This indirect method was adopted to lessen the possibility of other genetic material being transmitted at the same time as the K88 plasmid. As judged by ligated intestine tests in pigs, all 4 forms of the O141 strain produced the same amount of enterotoxin.

Table 3

The effect of giving day-old piglets, by mouth, forms of the enteropathogenic
E. coli *O141 : K85ab, 88ab strain possessing different combinations of*
the Hly and K88 plasmids

Form of strain	Number of piglets		
	to which it was given	that developed diarrhoea	that died†
Hly$^+$ K88$^+$	10	10	5
Hly$^-$ K88$^+$	10	9	5
Hly$^+$ K88$^-$	13	0	0
Hly$^+$ K88$^-$ K88$^+$*	10	9	6

* The Hly$^+$ K88$^-$ form into which a K88ab plasmid had been introduced from another strain of *E. coli.* All forms were Ent$^+$.
† or were killed when severely ill.
(From Smith & Linggood, 1971*b*.)

The incidence and severity of diarrhoea in the piglets given the Hly$^-$ K88$^+$ form of the O141 strain equalled that produced by the wild strain itself, i.e. the Hly$^+$ K88$^+$ form. By contrast, the piglets given the Hly$^+$ K88$^-$ form remained normal. The disease in the piglets given the Hly$^+$ K88$^-$ K88$^+$ form, i.e. the Hly$^+$ K88$^-$ form into which a K88 plasmid had been introduced, resembled that in the piglets given the wild strain itself. The results of estimating the numbers of *E. coli* in some of the piglets is summarized in Table 4. Very high concentrations of the infecting organisms were found in the chyme in the small intestine of the piglets given the Hly$^-$ K88$^+$ or the Hly$^+$ K88$^-$ K88$^+$ form and very low concentrations in this region in the piglets given the Hly$^+$ K88$^-$ form. No *E. coli* were found in the liver or spleen of these 4 piglets or in these organs of any animals used in subsequent experiments.

Table 4
The concentration of the infecting organisms in different parts of the alimentary tract of piglets given, by mouth, forms of the E. coli *O141 : K85ab 88ab strain possessing different combinations of the Hly and K88 plasmids*

| Organ | Log_{10} number of infecting organisms/g of content of stated organ, in a piglet given the | | | |
	$\text{Hly}^+\text{K88}^-$ form	$\text{Hly}^+\text{K88}^-$ form	$\text{Hly}^-\text{K88}^+$ form	$\text{Hly}^+\text{K88}^-\text{K88}^+$ form
Stomach	3.4	< 3.0	...	< 3.0
Small intestine				
portion 1	4.0	5.0(6.4)*	9.1	8.3
3	3.7	4.8(6.3)	9.3	7.9
5	5.0	5.3(6.7)	...	8.0
7	6.0	5.8(7.3)	...	8.6
Colon	8.0	7.3(8.7)	...	8.3

* The figures in parentheses are those for an $\text{Hly}^+\text{K88}^+$ form of the O141 strain which this piglet had acquired naturally from other litter-mate piglets that had been given it. The pig given the $\text{Hly}^+\text{K88}^-\text{K88}^+$ and the one given the $\text{Hly}^-\text{K88}^+$ form had severe diarrhoea; the other two were normal.
... = no observation; for other details see Tables 1 and 3.
(From Smith & Linggood, 1971*b*.)

When the different forms of the O141 strain were given to pigs, instead of piglets, the results were essentially the same (Table 5). A large group was given the $\text{Hly}^-\text{K88}^+$ form in the hope of discovering whether the Hly plasmid was playing a part in the pathogenesis of the disease, but no detectable difference was found between this group and the large group given the $\text{Hly}^+\text{K88}^+$ form. Two pigs were given mixtures of equal amounts of the $\text{Hly}^+\text{K88}^+$ and $\text{Hly}^-\text{K88}^+$ forms. Both developed diarrhoea; in every part of their alimentary tracts, the concentrations of the two forms were approximately the same (Table 6).

Table 5
The effect of giving weaned 8-week-old pigs, by mouth, forms of the E. coli *O141 : K85ab, 88ab strain possessing different combinations of the Hly and K88 plasmids*

| Form of strain | Number of pigs | | |
	to which it was given	that developed diarrhoea	that died
$\text{Hly}^+\text{K88}^+$	12	12	4
$\text{Hly}^-\text{K88}^+$	18	16	4
$\text{Hly}^+\text{K88}^-$	2	0	0
$\text{Hly}^+\text{K88}^-\text{K88}^+$	2	2	0

For other details see Table 3.
(From Smith & Linggood, 1971*b*.)

Table 6

The concentration of Hly$^+$ and Hly$^-$ O141 : K85ab, 88ab, organisms in different parts of the alimentary tract of a weaned 8-week-old pig given equal numbers of each by mouth

Organ	Log$_{10}$ number of infecting organisms/g of content of stated organ that were	
	Hly$^+$	Hly$^-$
Stomach	< 2.7	< 2.7
Small intestine		
portion 1	5.4	5.6
3	6.3	6.3
5	8.5	8.7
7	9.3	9.4
Colon	8.5	8.3
Rectum	8.5	8.3

This pig had diarrhoea when killed.
(From Smith & Linggood, 1971*b*.)

A final attempt to demonstrate a difference between Hly$^+$ and Hly$^-$ forms was made by infecting pigs with two such forms of a wild enteropathogenic O141 : K85ac strain, a strain which commonly produces oedema disease in addition to diarrhoea (Smith & Halls, 1968*b*); an Hly$^-$ form to which an Hly plasmid had been transmitted from another strain was included in this investigation. No difference was noted between the ability of these 3 forms to produce oedema disease or diarrhoea (Table 7).

(b) *The significance of the Ent and K88 plasmids in the ability of an O8 : K40, 88ab : H9 strain of* E. coli *to infect piglets*

This non-haemolytic strain had been isolated in abnormally high concentrations in the small intestine of a piglet suffering from a relatively mild attack of diarrhoea. It was unusual since it failed to produce a positive reaction when tested in the ligated intestine of pigs and piglets, i.e. it was Ent$^-$. The results of introducing an Ent plasmid into this strain and removing the K88 plasmid from it are summarized in Table 8. The diarrhoea that occurred in 3 of the 8 piglets given the K88$^+$ Ent$^-$ form was milder than that which occurred in the piglets given the K88$^+$ Ent$^+$ form. The results of bacteriological examinations on 3 piglets from one litter that had been given one or other of the 3 forms of the O8 strain are summarized in Table 9. Of the 2 piglets given forms containing the K88 antigen very high concentrations of the infecting organisms were found in the small intestine but not in the piglet given the K88$^-$ Ent$^+$ form.

Table 7

The effect of giving weaned 8-week-old pigs, by mouth, Hly⁺ and Hly⁻ forms of an Ent⁺ E. coli *O141 : K85ac strain*

Form of strain	Number of pigs		
	to which it was given	that developed diarrhoea	that developed oedema disease
Hly⁺	5	5	3
Hly⁻	5	3	2
Hly⁻ Hly⁺ *	6	6	3

* The Hly⁻ form into which an Hly plasmid had been introduced from another strain of *E. coli*. (From Smith & Linggood, 1971*b*.)

Table 8

The effect of giving piglets, by mouth, forms of an E. coli *O8 : K40, 88ab : H9 strain possessing different combinations of the K88 and Ent plasmids*

Form of strain	Number of piglets	
	to which it was given	that developed diarrhoea
K88⁺ Ent⁻	9	3
K88⁺ Ent⁺	9	8
K88⁻ Ent⁺	5	0

(From Smith & Linggood, 1971*b*.)

Table 9

The concentration of infecting organisms in different parts of the alimentary tract of piglets given, by mouth, forms of the E. coli *O8 : K40, 88ab : H9 strain possessing different combinations of the K88 and Ent plasmids*

Organ	Log_{10} number of infecting organisms/g content of stated organ, in a piglet given the		
	K88⁺ Ent⁻ form	K88⁺ Ent⁺ form	K88⁻ Ent⁺ form
Stomach	5.7	6.3	4.8
Small intestine			
portion 1	8.3	9.3	4.4
3	9.3	9.3	5.6
5	9.8	9.8	5.6
7	9.3	10.2	6.0
Colon	9.6	9.9	6.9

For further details see Table 1.
The piglet given the K88⁻ Ent⁺ form was normal. The other two piglets had diarrhoea; it was severest in the one given the K88⁺ Ent⁺ form.
(From Smith & Linggood, 1971*b*.)

(c) *The introduction of K88 and Ent plasmids into a non-pathogenic*
O9 : K36 : H19 strain of E. coli: *its consequent conversion to*
enteropathogenicity for piglets

This Hly⁻ Ent⁻ K88⁻ strain had been isolated from the faeces of a healthy pig. It neither dilated ligated intestinal segments of piglets nor caused clinical disease in them following oral administration. The results of testing, in piglets, forms of this strain that had acquired different combinations of K88 and Ent plasmids during mixed culture with enteropathogenic pig strains are summarized in

Table 10

The effect of giving piglets, by mouth, forms of an E. coli *O9 : K36 : H19*
strain possessing different combinations of the K88 and Ent plasmids

Form of strain*	Number of piglets	
	to which it was given	that developed diarrhoea
K88⁻ Ent⁻	8	0
K88⁻ Ent⁺	6	0
K88⁺ Ent⁻	11	3
K88⁺ Ent⁺	16	12

* All forms possessed plasmids for haemolysin production and colicine production and for resistance to neomycin, these plasmids being present in the strain that donated the K88 plasmid.
(From Smith & Linggood, 1971*b*.)

Table 11

The concentration of the infecting organisms in different parts of the alimentary
tract of piglets given, by mouth, forms of the E. coli *O9 : K36 : H19 strain*
possessing different combinations of the K88 and Ent plasmids

Organ	Log_{10} number of infecting organisms/g content of stated organ, in a piglet given the			
	K88⁻ Ent⁻ form	K88⁺ Ent⁻ form (1)*	K88⁺ Ent⁻ form (2)*	K88⁺ Ent⁺ form
Stomach	5.3	4.0	3.4	5.7
Small intestine				
portion 1	5.3	8.3	5.2	9.5
3	6.0	8.6	8.7	9.3
5	4.7	8.5	8.3	9.2
7	7.4	10.2	9.2	9.2
Colon	8.4	9.7	9.2	9.5

For further details see Tables 1 and 3.
* The results for 2 piglets given the K88⁺ Ent⁻ form are shown; piglet (1) was ill, but never had diarrhoea whereas piglet (2), like the one given the K88⁺ Ent⁺ form, had diarrhoea.
(From Smith & Linggood, 1971*b*.)

Table 10. Eighteen of the piglets were submitted to a bacteriological examination; the results for 4 of them are illustrated in Table 11. All of the piglets given K88⁻ forms of the 09 strain, irrespective of whether or not they possessed the Ent plasmid remained healthy and only low concentrations of the infecting organisms were found in their small intestines. By contrast, diarrhoea only occurred in piglets given K88⁺ forms but was more frequent and more severe in those given forms that were also Ent⁺. The characteristic bacteriological feature in piglets given K88⁺ organisms was their high concentration in the small intestines, the extent of proliferation being independent of their Ent status.

The relationship between possession of K88 antigen and proliferative ability was also demonstrated in experiments in which K88⁺ organisms of the 09 strain were mixed with 100 times their amount of K88⁻ organisms of this strain and given to piglets by mouth. These piglets were colostrum-deprived and, as a consequence, were largely free of antibodies. When the piglets were killed 24 h later, the ratio of K88⁺ to K88⁻ organisms in the stomach was approximately the same as that in the infecting dose, i.e. 1 to 100. In the small intestines, however, the K88⁺ organisms had achieved dominance, the ratio now being of the order of 10–1000 to 1. (Differentiation of these 2 kinds of organisms from each other was achieved by virtue of the fact that the K88⁺ ones were streptomycin-resistant and the K88⁻ ones were nalidixic acid-resistant.)

(d) *The effect of the K88 and Ent plasmids on the virulence of* Salmonella typhimurium *and* Salm. cholerae-suis

(i) *The K88 plasmid*

The results of infecting mice with *Salmonella typhimurium* and *Salm. cholerae-suis* with or without a K88ac plasmid from an enteropathogenic pig strain of *E. coli* are summarized in Tables 12 and 13. In both organisms the presence of the K88 plasmid was associated with reduced virulence. Strains from which the K88 plasmid had been removed were included in these studies to counteract the possible criticism that the organisms that had acquired the K88 plasmid during the mixed culture procedures might have been less virulent than the majority of the organisms of the parent strain. The K88⁺ and K88⁻ forms of the *Salm. cholerae-suis* strain were each given orally to 20 pigs. Although none of the pigs died, they all became ill, the course of the disease being rather more severe in the pigs given the K88⁻ form than in those given the K88⁺ form.

Bacteriological examination of mice killed at daily intervals after oral infection with K88⁺ or K88⁻ forms of the *Salm. typhimurium* strain revealed that, contrary to the findings in the *E. coli* infections of piglets, very much higher concentrations of salmonellae were found in the small intestines of mice

Table 12
The effect on virulence of introducing a K88 plasmid into a strain of Salm. typhimurium

Route of administration	Dose (viable organisms)	Form of strain	Number of mice	
			to which it was given	that died
Oral	10^9	K88⁻	25	6
		K88⁺	25	0
		K88⁻ (K88⁺)†	25	6
Intraperitoneal	10^3	K88⁺	15	2*
		K88⁻ (K88⁺)	15	12

* The *Salm. typhimurium* organisms isolated from the organs of these two mice had lost their K88 plasmid.
† The K88⁺ form from which the K88 plasmid had been removed.

Table 13
The effect on virulence of introducing a K88 plasmid into a strain of Salm. cholerae-suis

Form of strain	Number of mice	
	to which it was given	that died
K88⁺	25	4
K88⁻ (K88⁺)*	25	16

* The K88⁺ form from which the K88 plasmid had been removed.
Each mouse was given 10^9 viable organisms orally.

given the K88⁻ form than those given the K88⁺ form; examinations of the internal organs yielded similar results.

(ii) *The Ent plasmid*

When forms of a *Salm. typhimurium* strain containing an Ent plasmid from one of 3 enteropathogenic pig strains of *E. coli* were administered orally to groups of mice (Smith & Halls, 1968a), the mortality pattern in the 3 groups was no different from that in a similar group given the Ent⁻ parent strain. Diarrhoea was no more common in the Ent⁺ groups than in the Ent⁻ group. Similar results were obtained when the Ent⁺ and Ent⁻ forms were given orally to young calves. The virulence of *Salm. cholerae-suis* for pigs, mice and rabbits was also not altered as a result of accepting Ent. In all these salmonella virulence tests, the Ent plasmids involved were those coding for ST (heat-stable enterotoxin) only; it had not been possible to introduce into salmonellae the kind of plasmid that coded for ST and LT (heat-labile enterotoxin).

(e) *The effect of the K88 and Hly plasmids on the*
virulence of E. coli *for mice*

The results of giving intraperitoneally to groups of mice those forms of the enteropathogenic O141 : K85ab, 88ab strain possessing different combinations of the K88 and Hly plasmids are illustrated in Table 14. The Hly plasmid

Table 14

The effect of injecting intraperitoneally into mice forms of the O141 : K85ab,
88ab strain of E. coli *possessing different combinations of the*
Hly and K88 plasmids

Dose of viable organisms ($\times 10^6$)	Number of mice that died/number of mice given the			
	$Hly^+ K88^+$ form	$Hly^- K88^+$ form	$Hly^+ K88^-$ form	$Hly^- K88^-$ form
3	...	...	0/5	...
5	...	...	2/5	...
10	3/10	0/10	10/10	0/10
30	6/10	1/20	...	11/20
100	5/5	8/10	...	10/10

increased the virulence of the organisms possessing it and the K88 plasmid decreased it. This was in contrast to the behaviour of the same strain when given orally to piglets (Section 6(a)). There the Hly plasmid appeared to play no part in the development of the disease whereas the possession of the K88 plasmid was essential.

(f) *The effect of the K99 and the K88 plasmids on the ability of*
an enteropathogenic O8 : K85,99 strain of E. coli *to infect lambs*

Twin lambs, one infected orally with a wild O8 : K85,99$^+$ strain of *E. coli* enteropathogenic for calves and lambs and the other infected with a K99$^-$ form of this strain, are illustrated in Table 15. The results for a lamb given a K99$^-$ K88$^+$ form, i.e. the K99$^-$ form into which a K88 plasmid had been introduced from a pig enteropathogenic strain, are included. Only the lamb given the wild K99$^+$ form developed diarrhoea and it had much higher concentrations of the infecting strain in its small intestine than the other 2, the difference being particularly noticeable in the lower small intestine. However, concentration in the upper intestine of the diarrhoeic lamb was much lower than that usually found in lambs given the same strain. The avirulent nature of the K99$^-$ form of the O8 strain was further demonstrated by its failure to produce diarrhoea in another 7 lambs. Despite the fact that it produced enterotoxin active on piglet

Table 15

The concentration of the infecting organisms in different parts of the alimentary tract of lambs given, by mouth, different forms of an enteropathogenic O8 : K85, 99 strain of E. coli

Organ	Number of organisms, $\log_{10}/g.$, in stated organ, in a lamb given a		
	$K99^+$ form	$K99^-$ form	$K99^- K88^+$ form*
Stomach	6.5	5.3	5.7
Small intestine			
portion 1	5.7	5.4	5.3
3	6.6	5.4	6.0
5	9.0	5.2	6.0
7	9.3	7.9	6.4
Colon	9.5	8.5	8.6

The lamb given the $K99^+$ form, i.e., the wild strain, had diarrhoea; the other two were normal when killed.

* The $K99^-$ strain with a K88 plasmid introduced from an enteropathogenic pig strain.

For other details see Table 1.

intestine, the $K99^+$ form failed to produce diarrhoea in piglets on oral inoculation.

7. Conclusions

The results of the infection experiments in piglets clearly demonstrated that possession of the K88 plasmid was very involved in the ability of *E. coli* organisms to proliferate in the anterior small intestine. Strong evidence supporting the view that they actually do so by adhering to the epithelium was obtained in fluorescent-microscopy studies by Jones & Rutter (1972) working with $K88^+$ and $K88^-$ forms of the same pig enteropathogenic strain of *E. coli*. Some pig enteropathogenic strains, however, do not possess the K88 antigen, e.g. the O141 : K85ac strain that produced both diarrhoea and oedema disease (Section 6(a)). These strains proliferate well in the anterior small intestine and there is some evidence that they too, do so by adhering to the intestinal epithelium (Smith & Halls, 1968). These strains may produce other substances that function in the same manner as the K88 antigen. Calf and lamb enteropathogenic strains certainly appear to produce such a substance, the substance in their case being the plasmid-controlled K99 antigen, K99 adhesion being specific for the calf and lamb and K88 adhesion specific for the pig. In the salmonella, and mouse intraperitoneal infections with the enteropathogenic O141 : K85ab, 88ab strain of *E. coli*, K88 organisms were always at a disadvantage compared with $K88^-$ organisms. It is only in infections of the porcine alimentary tract that possession of the K88 plasmid by the infecting organisms was of paramount importance.

The Ent plasmid only became important in the pathogenesis of *E. coli* infection when the organisms were already adhering to the small intestinal epithelium. It was responsible for the production of enterotoxin which converted a mild, or non-existent, diarrhoea into a severe and, perhaps, fatal one.

The significance of haemolysin production in porcine diarrhoea remains a mystery — although the presence of the Hly plasmid was associated with increased virulence in intraperitoneally-induced infections of mice, no significance could be attached to it in oral infections of pigs. Why then do so many strains of *E. coli* enteropathogenic for pigs produce haemolysin? The answer to this question may lie in the fact that plasmids, like viruses, are autonomous. Consequently, we should, perhaps, be thinking more about the survival of Hly plasmids themselves rather than of the bacterial cells they inhabit.

The present investigation serves as a demonstration that plasmids can be utilized to study subjects other than the more obvious ones of bacterial genetics and antibiotic resistance. It is possible, therefore, that in future they will be exploited to a greater extent and for a variety of purposes.

8. References

JONES, G. W. & RUTTER, J. M. (1972). Role of K88 antigen in the pathogenesis of neonatal diarrhoea caused by *Escherichia coli* in piglets. *Infect. & Immunity* **6**, 918.

ORSKOV, I. & ORSKOV, F. (1966). Episome-carried surface antigen K88 of *Escherichia coli*. 1. Transmission of the determinant of the K88 antigen and influence on the transfer of chromosomal markers. *J. Bact.* **91**, 69.

SKERMAN, F. J., FORMAL, S. B. & FALKOW, S. (1972). Plasmid-associated enterotoxin production in a strain of *Escherichia coli* isolated from humans. *Infect. & Immunity* **5**, 622.

SMITH, H. W. (1971). The production of diarrhoea in baby rabbits by the oral administration of cell-free preparations of enteropathogenic *Escherichia coli* and *Vibrio cholerae:* the effect of antiserum. *J. med. Microbiol.* **5**, 299.

SMITH, H. W. & GYLES, C. L. (1970). The relationship between two apparently different enterotoxins produced by enteropathogenic strains of *Escherichia coli* of porcine origin. *J. med. Microbiol.* **3**, 387.

SMITH, H. W. & HALLS, S. (1967*a*). The transmissible nature of the genetic factor in *Escherichia coli* that controls haemolysin production. *J. gen Microbiol.* **47**, 153.

SMITH, H. W. & HALLS, S. (1967*b*). Observations by the ligated intestinal segment and oral inoculation methods on *Escherichia coli* infections in pigs, calves, lambs and rabbits. *J. Path. Bact.* **93**, 499.

SMITH, H. W. & HALLS, S. (1968*a*). The transmissible nature of the genetic factor in *Escherichia coli* that controls enterotoxin production. *J. gen. Microbiol.* **52**, 319.

SMITH, H. W. & HALLS, S. (1968*b*). The production of oedema disease and diarrhoea in weaned pigs by the oral administration of *Escherichia coli:* factors that influence the course of the experimental disease. *J. med. Microbiol.* **1**, 45.

SMITH, H. W. & JONES, J. E. T. (1963). Observations on the alimentary tract and its bacterial flora in healthy and diseased pigs. *J. Path. Bact.* **86**, 387.

SMITH, H. W. & LINGGOOD, M. A. (1971*a*). The transmissible nature of enterotoxin in a human enteropathogenic strain of *Escherichia coli*. *J. med. Microbiol.* **4**, 301.

SMITH, H. W. & LINGGOOD, M. A. (1971*b*). Observations on the pathogenic properties of the K88, Hly and Ent plasmids of *Escherichia coli* with particular reference to porcine diarrhoea. *J. med. Microbiol.* **4**, 467.

SMITH, H. W. & LINGGOOD, M. A. (1972). Further observations on *Escherichia coli* enterotoxins with particular regard to those produced by atypical piglet strains and by calf and lamb strains: the transmissible nature of these enterotoxins and of a K antigen possessed by calf and lamb strains. *J. med. Microbiol.* **5**, 243.

SOJKA, W. J. (1971). Enteric diseases in new-born piglets, calves and lambs due to *Escherichia coli* infection. *Vet. Bull.* **41**, 509.

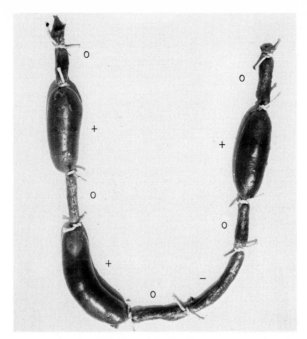

Plate 1. Ligated segments of pig intestine inoculated with cultures of *E. coli*. The ligatures are extra ones applied *post mortem* to show the position of the real ones. +, dilated and −, undilated segments that had been inoculated respectively with, enteropathogenic and non-enteropathogenic strains. 0, uninoculated segments. × 0.25 (from Smith & Halls, 1967*b*).

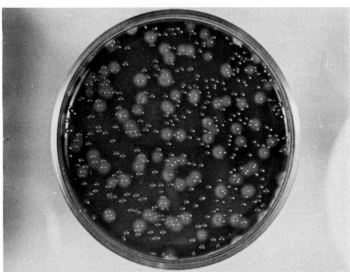

Plate 2. Result of a mixed culture experiment. The mixed culture of a haemolytic donor strain of *E. coli* and a non-haemolytic recipient strain of *E. coli* has been inoculated on to a washed blood agar plate containing an antibiotic which has completely suppressed the growth of the donor strain. The colonies are those of the antibiotic-resistant recipient strain; *c*. 10% of them are now haemolytic. × 0.75 (from Smith & Halls, 1967*a*).

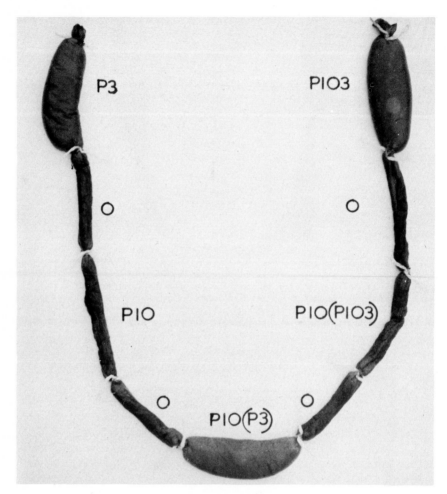

Plate 3. Ligated segments of pig intestine inoculated with cultures of *E. coli*. The segments are identified according to the strains used for inoculating them. P3 and P103, enterotoxigenic strains. P10, non-enterotoxigenic strain. P10(P3) and P10(P103), passaged mating cultures of P10 in which P3 and P103 had been employed as prospective donors. The ability to produce enterotoxin has been transferred to P10 from P3 but not from P103. 0, uninoculated segment. × 0.375 (from Smith & Halls, 1968*a*).

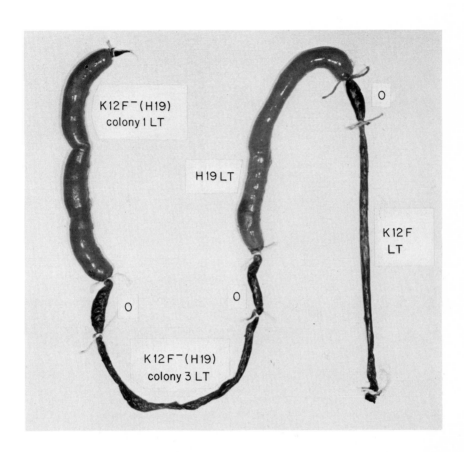

Plate 4. Ligated segments of rabbit intestine after injection of LT preparations of *E. coli*. The segments are identified according to the materials injected into them. K12F⁻ (H19) Ent⁺ colony 1 and K12F⁻ (H19) Ent⁻ colony 3 are 2 colonies of recipient strain K12F⁻ that did or did not receive the Ent plasmid from a human enteropathogen *E. coli* strain H19 (026 : K60) during mixed culture. 0, uninoculated segments. × 0.75 (from Smith & Linggood, 1971a).

Plate 5. The baby rabbit on the left had been given, 20 h previously, by mouth, 5 ml of an LT preparation of *E. coli* K12 in which had been established the Ent plasmid of a pig enteropathogenic O8 : K87, 88ab *E. coli* strain. The one on the right had been given a similar preparation of a K12 Ent⁻ strain at the same time. × 0.9 (from Smith, 1971).

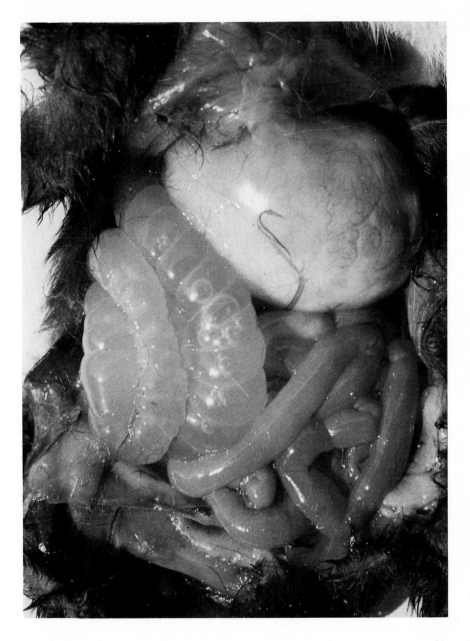

Plate 6. This baby rabbit had been given, 20 h previously, by mouth, 5 ml of an LT preparation of *E. coli* K12 in which had been established the Ent plasmid of a pig enteropathogenic O8 : K87, 88ab *E. coli* strain. The small intestine in addition to the large intestine contained large amounts of watery fluid. × 3.75 (from Smith, 1971).

Mycotoxins in Animal Feeds

A. HACKING AND J. HARRISON

*A.D.A.S., Ministry of Agriculture, Fisheries and Food,
Shardlow Hall, Shardlow, Derby, England*

CONTENTS

1. History

MOULDS, as invaders of animal tissue were first recorded by De Réaumur (1749). Later, Montagne (1813) reported 'mould or blue mucor' in the air sac of a duck and Rayer & Montagne (1842) identified an *Aspergillus* invading the air sac of a bullfinch. A lung lesion in deer reported by Rousseau & Serrurier (1841), was the first recorded mammalian instance, while Virchow (1856) was the first to report such invasion in man. About this time, the discovery of bacterial invasion as the cause of disease changed medical practice and overshadowed the rarer cases of reported mycoses, physical invasion of the host tissue by the growing mould. More recently, studies in mycoses have increased and numerous mycoses leading generally to death have been reported (Ainsworth & Austwick, 1973). One such case in poultry has been investigated in the field (Hacking & Blandford, 1971).

This introduction has referred to moulds causing mycoses; On the other hand, mycotoxins, as metabolites of mould growth, can cause clinical conditions in the absence of the mould concerned. Moulds have been present from time immemorial where man has stored food either for himself or his animals. If moisture and temperature in the store are optimum for mould proliferation toxic metabolites may be formed. All metabolites are not toxic, a statement no doubt endorsed by 'Blue' cheese connoisseurs.

With the discovery of aflatoxin (Allcroft & Carnaghan, 1963) and the subsequent research interest in 'mycotoxins', the impression might be given that this is a completely new area of work. In medieval history references to St. Anthony's Fire, places the subject back in historical perspective. As early as the mid-16th century a link had been established between St. Anthony's Fire and 'scabrous grain'. The active principle from ergot, the sclerotium of *Claviceps purpurea,* was not known until the early 1930s, though crude extracts of sclerotia had been used for over a century in medicine.

Another clinical condition associated with the mycological invasion of feed is Alimentary Toxic Aleukia (ATA). This condition, known in Russia for over a century, reached epidemic proportions in Western Siberia in 1932. The grain involved was contaminated with *Fusarium* species, and a full description of the chemical state is recorded by Joffe (1971).

Biochemical extraction techniques for mycotoxins were developed from research into fungal metabolites as producers of antibiotics. In some cases, though toxicity was recorded against test organisms, similar toxicity was observed in the test animals. Such a compound was patulin, described by Waksman in 1942-43 at a Therapeutic Research Corporation Committee as 'an excellent rat poison'.

Thus, from the knowledge of ergot poisoning and ATA and the basic biochemistry of antibiotic research, mycotoxin research had a good base when aflatoxin toxicity occurred in turkey flocks.

2. Changes in Agriculture

It is almost certain that clinical conditions due to mycotoxins have occurred sporadically in livestock for centuries. Intensification of farming, changes in storage conditions and the handling of cereals in bulk have made the problem potentially greater (Harrison & Hacking, 1974). Mould spores present in material can proliferate when environmental conditions are suitable producing, occasionally, deleterious metabolites. The presence of any particular organism is only part of the story. Following the outbreak of aflatoxicosis in turkeys, strains of the causal organism *Aspergillus flavus,* were detected in home-grown cereals stored moist in grain silos, without aflatoxin production. The theory was propounded that British, unlike tropical, strains were incapable of toxin production. This was disproved when maize or groundnut instead of barley was used as substrate for local isolates. The higher lipid and carbohydrate content of the former substrates greatly favoured aflatoxin production.

Preliminary information from a survey of *Fusarium* strains on barley destined for feeding to livestock in the United Kingdom has shown that 20–30% of isolates produce zearalenone (F-2) when grown on a suitable substrate.

3. Case Histories

As a guide to what might constitute an incidence of mycotoxicosis in the field, the simple criteria proposed by Forgacs & Carll (1962) can be followed. These criteria are: (1) the disease or illness is neither infectious nor contagious and is generally confined to a single farm; (2) the occurrence of the condition is often associated with a particular feed or change of feed; (3) treatment with drugs and antibiotics is ineffective; (4) pathological examination does not detect causal bacteria or viruses; (5) though some symptoms are similar to avitaminosis, vitamin treatment is ineffective; (6) moulds are present in the feed. The natural extension of this statement of criteria is that the detection of a chemical agent which is known to produce the distinct clinical effects, is required.

Other reviews (Borker *et al.*, 1966; Brook & White, 1966; Feuell, 1966; Mirocha, Christensen & Nelson, 1968, 1969; Mirocha *et al.*, 1968; Wilson, 1968) have categorized the main groups of toxins broadly as nephrotoxins, hepato-toxins, neurotoxins, haematological effects and those responsible for general digestive disorders.

At the ADAS/MAFF laboratories, Shardlow, circumstantial evidence indicates that a number of mycotoxins occurs sporadically in the U.K. Citrinin, a nephrotoxin recorded by Krogh *et al.* (1973) was detected in a sample of oats associated with a diuretic effect when fed to racehorses.

Aflatoxin, the major hepatotoxin, has been detected in various feeds examined. This mycotoxin is the only one for which specific standard levels in feedstuffs have been laid down.

A neurotoxin also detected at Shardlow was cyclopiazonic acid from a feed which was implicated in a nervous uncoordination of the hind quarters in calves.

Haematological effects can be of 2 types: (1) Constriction of blood vessels as with ergot toxicity, which leads to gangrene of the extremities in sheep and cattle and the feet, wattles, etc., of poultry. Gangrene was observed in poultry in a case recorded at Shardlow. (2) Direct effect on white blood cells with subsequent haemorrhage in the bowel. This condition, classed clinically as a type of haemorrhagic bowel syndrome, has tentatively been attributed to the tricothecin group of toxins (Ueno *et al.*, 1973) produced by members of the genus *Fusarium*. It is considered as a possible mechanism by which ATA occurs.

At the Edinburgh School of Agriculture, feed inoculated with *Fusarium tricinctum* a tricothecin producer, has been fed to pigs and observations made on their blood cells. A distinct reduction in white cells was observed, though this did not lead to the clinical bowel syndrome condition (unpublished data). Further work is to continue on pure tricothecin in an attempt to clarify these initial observations.

The main interest at Shardlow has been the oestrogen zearalenone (F-2) produced by *F. graminearum* and *F. culmorum*. The interest in this metabolite began with a case of infertility in a dairy herd in 1967. The work of C. J. Mirocha in the United States had shown that this metabolite produced by *F. graminearum* in maize had been associated with infertility problems in pigs (Fig. 1). Infertility in the dairy herd was attributed to the hay in the diet, by a process of elimination of the other constituents of the feed (Mirocha *et al.*,

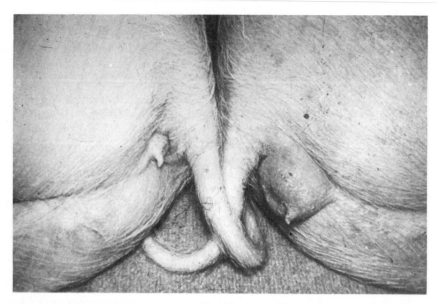

Fig. 1. Vulvo-vaginitis in pigs. L, normal pig; R, pig showing the clinical effect of zearalenone (F-2) produced by *Fusarium* spp.

1968). Biochemical analyses showed the metabolite zearalenone to be present.

In the winter of 1967/68 poultry flocks in the north west of England showed a decrease in egg production. Analysis of the whole feed showed zearalenone to be present and analysis of individual components of the feed showed that barley was the prime constituent responsible for its formation. At the same time as this investigation was being carried out, 2 incidents of reduced litter size in pigs were brought to the attention of our laboratory. The normal litter size of 10–12 was reduced drastically to 4–5 when certain barley grain was incorporated into the diet. In both incidents the presence of zearalenone was confirmed in the suspect grain samples. The above circumstantial evidence from case histories in both poultry and pigs made it essential to make a more critical evaluation of the mycotoxins by feeding trials. Techniques for the production of purified zearalenone were not available at the time of the first feeding trials. These initial trials, made with the kind collaboration of various feed compounders, used grain infected deliberately with a known F-2 producing strain of *Fusarium*. The grain was incubated until thin layer chromatography indicated an appreciable concentration of the toxin. This barley was then incorporated into the diet of the flock and assessed against a diet containing non-inoculated barley.

Twenty-four laying hens were monitored for 3 weeks prior to incorporation of this feed in the diet in order to establish a standard egg-laying pattern. A marked reduction in egg-laying ability was recorded after the incorporation of

Fig. 2. Splay leg in pigs; the clinical effect of zearalenone (F-2) on pig litters.

the infected grain in the feed. This response was detected in the first 2 days of the experiment and was followed by a rate of lay well above normal. Continued feeding showed a further reduction in egg-laying ability to *c.* 1/5th of normal. However, the supply of infected grain ran out and the laying pattern returned to normal. Ovaries and oviducts were removed from the birds at the end of the trial for histological examination to ascertain whether the oestrogenic material might have affected the reproductive tract. Though abnormalities were detected these were thought to be attributed to an earlier infection with infectious bronchitis.

Published techniques by Shotwell, Hesseltine & Goulden (1969) and Sherwood & Peberdy (1972) were then adopted at the Shardlow laboratories and enabled us to increase the yield, purity and accuracy of assay of zearalenone. A further trial was conducted with the pure material incorporated into the diet at the level of 50 mg/kg liveweight of the birds. No effect on egg production was observed but it was considered possible that in the purification of the zearalenone another associated metabolite which could have been responsible for the initial fall in egg production effect in the first trial, had been discarded.

A further field study involving the presence of zearalenone in a barley diet fed to pregnant sows was brought to our attention. In this case zearalenone at a level of 75 mg/kg bodyweight of the affected animals was detected. The clinical condition was splay-leg in the litters and a reduction in litter size (Fig. 2). Feeding trials using both laboratory animals and pregnant sows and gilts, produced the clinical symptoms observed in the field (Miller *et al.*, 1973).

As has been mentioned previously the metabolite of *F. tricinctum,* trico-thecin, is thought to be the causative metabolite for the condition Alimentary Toxicaleukia. The description of this condition was similar to that of a haemorrhagic syndrome in pigs. Samples of feed where this clinical condition had been detected showed the presence of the mycotoxin tricothecin. In an attempt to confirm that this mycotoxin might be the causal or predisposing agent, further feeding trials were made. In these trials regular samples of blood were taken over a period of 18 weeks and the effect on the white cell counts observed. There was a distinct reduction in lymphocyte and leucocyte counts. This could affect the immunological status of the animal and reduce natural resistance. Further work is to continue on the effects of this particular metabolite on pigs.

4. Assay Procedures

Various bioassay methods have been used, the relative merits of which are reviewed. The 3 techniques of major significance are those using brine shrimp (Brown, Wildman & Eppley, 1968) chick embryo (Verrett, Marliac & McLaughlin, 1964) and reaction of the mycotoxin as an antibiotic against various test bacteria (Clements, 1968). These assays are, in the main, related specifically to the assay of aflatoxin. They are now being used in ADAS Microbiology Laboratories in association with modifications to the extraction procedures for the evaluation of possible mycotoxins, where suspected animal feeds are associated with clinical conditions in the field.

5. Biochemical Techniques

Most techniques are based on solvent extraction of the starting material followed by screening by thin layer chromatography and subsequent chemical con-firmation by gas liquid chromatography or UV spectrophotometry. The major problem with biochemical extracts is the isolation of the chemical required without too much extraneous material. Column chromatography, solvent separations and cleaning-up procedures in general can lead to the loss of important material. Similarly it is important that the feed under analysis is the one causing the problem, which has itself been clearly defined. The difficulties of sampling are problems in themselves.

6. Future Developments

If problems exist either in connection with the use of home-grown cereals or imported ingredients in diets, established, recognized clinical conditions must be defined. Causative organisms, metabolites and the dose levels responsible must be tested under controlled laboratory conditions. Correlation of biological and biochemical assay methods is essential. It may be possible in the future to detect mycotoxins in various tissues of infected animals as a standard procedure. The world-scale transport of feed materials makes many aspects of the work difficult.

As mentioned previously, the need for a truly representative sample is of prime importance. Feed rejected for human consumption, because of visual moulding, can find its way to the animal feedstuffs market, with possible deleterious effects. Metabolism in the animal can direct either the mycotoxin or an equally toxic metabolic product to meat or milk, where it can be ingested by humans at a later date (Schuller, Verhulsdonk & Pausch, 1973; Krogh *et al.*, 1974).

Also, the gross effects of moulds on feed need study. Over-emphasis on the harmful effect could blind workers to beneficial compounds. A substance toxic to one type of stock may have desirable attributes when fed in appropriate concentrations to another. Zearalenone (F-2) can be seen in this context since a derivative zearalanol is subject to patent rights in the U.S.A. as a food additive for promoting liveweight gain in livestock. Also there is potential improvement in food conversion (Andrews, 1965; Hodge *et al.*, 1966; Moran, 1972). Again in this context, the gross mycological effect on nutrient levels of protein carbohydrate and lipid are worthy of study.

If some of the thorny problem areas are to be truly resolved, then mycologists, biochemists, veterinarians and nutritionists will need to share their expertise. Only then will an assessment of the incidence and frequency of mycotoxin in the U.K. be fully evaluated in economic terms.

7. References

AINSWORTH, G. C. & AUSTWICK, P. K. C. (1973). *Fungal Diseases of Animals.* Review Series No. 6, Commonwealth Bureaux of Animal Health. Farnham Royal Slough, England: C.A.B.

ALLCROFT, R. & CARNAGHAN, R. B. A. (1963). Toxic products in groundnut. Biological effects. *Chemy Ind. (London)* p. 50.

ANDREWS, S. (1965). U.S. Pat 3196019 (to Purdue Research Foundation).

BORKER, E., INSALATA, N. F., LEVI, C. P. & WITZEMAN, J. S. (1966). Mycotoxins in feeds and foods. *Adv. appl. Microbiol.* 8, 315.

BROOK, P. J. & WHITE, E. P. (1966). Fungal toxins affecting mammals. *Ann. Rev. Phytopathol.* 4, 171.

BROWN, R. F., WILDMAN, J. D. & EPPLEY, R. M. (1968). Temperature dose relationships with aflatoxin on the brine shrimp, *Artemia salina. J. Ass. off. anal Chem.* 51, 905.

CLEMENTS, N. L. (1968). Rapid confirmatory test for aflatoxin B, using *Bacillus megaterium. J. Ass off. anal. Chem.* 51, 1192.

FEUELL, A. J. (1966). Toxic factors of mould origin. *Can. med. Ass. J.* 94, 574.

FORGACS, J. & CARLL, W. T. (1962). Mycotoxicoses. *Adv. vet. Sci.* 7, 273.

HACKING, A. & BLANDFORD, T. A. (1971). Aspergillosis in 5-16 week old turkeys. *Vet. Rec.* 88, 519.

HARRISON, J. & HACKING, A. (1974). The occurrence and effects of mycotoxins in animal feedstuffs in the United Kingdom. Agricultural Development and Advisory Service, Ministry of Agriculture, Fisheries & Food. *Science Arm Ann. Rep. for 1972,* p. 73. London: H.M.S.O.

HODGE *et al.* (1966). U.S. Pat 323934/2 (to Commercial Solvents Corp.).

JOFFE, A. Z. (1971). Alimentary toxic aleukia. In *Microbial Toxins.* Eds S. Kadis & A. Ciegler. New York: Academic Press.

KROGH, P., HALD, B. & PEDERSEN, J. (1973). Occurrence of Ochratoxin A and citrinin in cereals associated with mycotoxic porcine nephropathy. *Acta path. microbiol. scand.* **B. 81,** 689.

KROGH, P., AXELSEN, N. H., ELLING, F., GYRD-HANSEN, N., HALD, B., HYLDGAARD-JENSEN, J., LARSEN, A. E., MADSEN, A., MORTENSEN, H. P., MØLLER, T., PETERSEN, O. K., RAVNSKOV, U., ROSTGAARD, M. & AALUND, O. (1974). Experimental porcine nephropathy. *Acta. path. microbiol. scand.* A. Suppl. No. 246.

MILLER, J. K., HACKING, A., HARRISON, J. & GROSS, V. J. (1973). Stillbirths, neonatal mortality and small litters in pigs associated with the ingestion of *Fusarium* toxin to pregnant sows. *Vet. Rec.* **93,** 555.

MIROCHA, C. J., CHRISTENSEN, C. M. & NELSON, G. H. (1968). Physiologic activity of some fungal oestrogens produced by *Fusarium. Cancer Res.* **28,** 2319.

MIROCHA, C. J., HARRISON, J., NICHOLLS, A. A. & McCLINTOCK, M. (1968). Detection of a fungal oestrogen (F-2) in hay associated with infertility in dairy cattle. *Appl. Microbiol.* **16,** 797.

MIROCHA, C. J., CHRISTENSEN, C. M. & NELSON, G. H. (1969). Biosynthesis of the fungal oestrogen F-2 and a naturally occurring derivative (F-3) by *Fusarium moniliforme. Appl. Microbiol.* **17,** 482.

MORAN, J. B. (1972). The effect of zearalanol on seasonal growth rates of cattle in a dry monsoonal environment. *Aust. J. exp. Agric. Anim. Husb.* **12,** 345.

SCHULLER, P. L., VERHULSDONK, C. A. H. & PAUSCH, W. E. (1973). Analysis of aflatoxin M, in liquid and powdered milk. Control of mycotoxins. Symposium Göteborg 1972. London: Butterworths.

SHERWOOD, R. F. & PEBERDY, J. F. (1972). Factors affecting the production of zearalenone by *Fusarium graminearum* in grain. *J. Stored Product Res.* 8, 71.

SHOTWELL, O. L., HESSELTINE, C. W. & GOULDEN, M. L. (1969). Note on the natural occurrence of ochratoxin. *J. Ass. off. anal. Chem.* **52,** 81.

UENO, Y., SATO, N., ISHII, K., SAKAI, K., TSUNODA, H. & ENOMOTO, M. (1973). Biological and chemical detection of tricothecene mycotoxins of *Fusarium* species. *Appl. Microbiol.* **25,** 699.

VERRETT, M. J., MARLIAC, J. D. & McLAUGHLIN, J. (1964). Use of chick embryo in the assay of aflatoxin toxicity. *J. Ass. off. agric. Chem.* **47,** 1003.

WILSON, B. J. (1968). *Mycotoxins in the Safety of Food.* Ed. J. C. Ayres. Westpoint, Connecticut.

The references to DE RÉAUMUR (1749), MONTAGNE (1813), ROUSSEAU & SERRURIER (1841), RAYER & MONTAGNE (1842) and VIRCHOW (1856) are cited by:

AUSTWICK, P. K. C. (1965). Pathogenicity. In *The Genus* Aspergillus. Eds K. B. Raper & D. I. Fennell. Baltimore: The Williams & Wilkins Co.

Mycotoxins in Food

B. JARVIS

British Food Manufacturing Industries Research Association,
Randalls Road, Leatherhead, Surrey, England

CONTENTS

1. Introduction

EXAMINATION of recent scientific literature reveals an almost exponential increase in the number of publications relating to mycotoxins, the fungi which produce them, the conditions and substrates in which they are produced and the diseases associated with specific mycotoxins. The number of fungal metabolites now known to be toxic grows yearly, as also does the range of foods and feeding stuffs in which mycotoxins have been shown to occur. Mycotoxins are important to man, as well as to his livestock, since in high doses many produce acute disease syndromes whilst at lower doses many have carcinogenic, teratogenic or oestrogenic effects in experimental animals. Mycotoxins known, or believed, to be responsible for disease syndromes in man are listed in Table 1, together with others which may be of significance in the future.

Although mycotoxins are produced as a direct result of the metabolic processes of microfungi, it is not always necessary for mould growth to have occurred on a particular food for mycotoxins to be present. Mycotoxin residues in foods may arise by a number of different routes (Table 2). Growth of toxinogenic fungi on agricultural produce may occur either in the field or during post-harvest drying, storage or distribution. Even though stored agricultural produce may not be used directly for human consumption, its use in animal

Table 1

Mycotoxins of possible significance to human health

Toxin	Disease syndrome	Country	Evidence	Reference
Ergot alkaloids	Ergotism – St. John's Fire	UK, France, USA, USSR	+	Joffe (1971)
Fusarial toxins	Alimentary Toxic Aleukia (septic angina)	Russia, Ukraine	+	
Stachybotryotoxin	Dermal toxicity; leukopenia	Russia	+	Forgacs (1965)
Psoralens	Contact Dermatitis	U.S.A.	+	Perone, Scheel & Meitus (1964)
Aflatoxins	E.F.D.V. Syndrome	Thailand	±	Shank (1971)
	Reye's Disease	New Zealand Australia	∓	Becroft & Webster (1972)
	Primary hepatoma	Uganda	±	Crawford (1971)
		Phillipines	±	Campbell & Salamat (1971)
Luteoskyrin, citrinin and others	'Yellow rice' toxicity	Japan	±	Saito, *et al.*, (1971)
Nivalenol ⎱ Fusarenon X ⎰	Nausea, vomiting, drowsiness haemorrhage	Japan	±	Ueno (1971)
Fusarial toxins including zearaleone	Vomiting, nausea	Italy	?	E. Sottini (pers. comm., 1971)
Ochratoxin	?		?	–

+, confirmed; ±, circumstantial; ∓, possible.

feeds may cause mycotoxicosis in farm animals. At low levels of toxin ingestion, symptoms of mycotoxicosis may not be evident in the living animal, or even on superficial post-mortem inspection such as is applied in a slaughter house. Yet evidence is accumulating that mycotoxins originating in the feed may be deposited in the body tissues or excreted in the milk of lactating mammals.

Table 2

Possible routes for mycotoxin contamination of human foods

1.	Mould damaged foodstuffs		
	(a)	Agricultural products e.g.	cereals oilseeds fruits vegetables
	(b)	Consumer foods	
2.	Residues in animal tissues and animal products e.g.		meat milk dairy products
3.	Mould-ripened foods e.g.		cheeses meat products
4.	Fermentation products e.g.		microbial proteins enzymes other food additives

Growth of toxinogenic fungi on manufactured food products can give rise to mycotoxin contamination. It is by no means proven, however, that toxin formation occurs during 'normal' mould spoilage of foods stored for excessive periods of time under adverse conditions. The use of microfungi in food fermentations (e.g. blue cheeses, fermented sausages and hams) could result in the formation of mycotoxins in the food. Finally, fermentation products such as fungal protein, enzymes, antibiotics and acids could become contaminated with mycotoxins if the strain of organism used in their production were toxigenic.

The purpose of the present paper is to assess the evidence available for mycotoxin contamination of human foods and to discuss briefly the significance to the consumer. For reasons of brevity the emphasis will be placed on human foods in Western countries.

2. Mycotoxins in Stored Agricultural Products

Stored products, which may be damaged by growth of toxinogenic fungi, range from cereals (e.g. wheat, barley, oats, maize, rice, sorghum, etc.) and oilseeds (e.g. peanuts, hazel nuts, brazil nuts, pistacchios, almonds, etc.) to fruits and vegetables.

(a) *Cereals*

Contamination of cereals by toxins such as aflatoxins, ochratoxin and zearalenone has been reported frequently (reviewed by Hesseltine, 1973; Scott, 1973). Workers in the U.S.A. have shown that mycotoxin contamination is associated largely with low grade cereals which are not normally used for human consumption (Shotwell *et al.,* 1969). However, surveillance studies do not adequately cover all stored cereals and problems of sampling in relation to the non-homogeneous distribution of toxic grains within a large bulk of material makes difficult the interpretation of results from surveillance and routine screening procedures. Undoubtedly the only way to prevent mycotoxin formation is the maintenance of good husbandry and the adequate and rapid drying of cereals to a water content at which mould growth cannot occur. Even then problems can arise from moisture migration during storage.

In general it is unlikely that significant concentrations of mycotoxins appear in flour and other cereal products for human consumption in the developed countries. There is evidence that a different situation exists in developing countries and agricultural communities (Barnes, 1970; Crawford, 1971; Purchase, 1971). Growth of toxinogenic moulds on cereals used for brewing could occur during the malting process, but good industrial control will prevent this. However, if a mycotoxin (e.g. aflatoxin) is present in barley which is to be used for brewing it is unlikely that it would be destroyed by the fermentation process, although it might be largely diluted out in the final product (Campbell, 1972). If bread is made from aflatoxin-contaminated cereal there is no reduction in aflatoxin content during 'proving' of the bread but during baking the level falls to *c.* 50% of the initial level (Jemali & Lafont, 1972).

(b) *Nuts and oilseeds*

In recent years shipments of aflatoxin-contaminated peanuts, hazel nuts, pistacchios and other commodities, such as figs, have been rejected by authorities responsible for control of imported foodstuffs. Investigations with confectionery-grade nuts in this country have failed to reveal any evidence of aflatoxins in nuts selected by food manufacturers for processing or in samples taken from retail outlets (Jarvis & Denizel, unpublished data). Studies of Turkish and Iranian pistacchios have suggested that contamination of the nuts probably occurs after soaking to remove the hulls from the shells (Rahnema, 1972; Denizel, personal communication, 1974). However, under suitable conditions of storage, aflatoxinogenic strains of *Aspergillus flavus* and *A. parasiticus* gain entry to the nut along the vascular system by which the kernel is attached to the shell. Significant levels of aflatoxin may occur in the kernel without obvious evidence of growth of *A. flavus* on the outside of the shell (Denizel, pers. comm., 1974).

Major users of confectionery-grade nuts are well aware of the problems associated with aflatoxin contamination. Manufacturers carry out testing and selection procedures at all stages from import of the nuts to final product assessment. Consignments of nuts which are rejected for human use are used either for animal feed compounding or for oil extraction. Aflatoxin residues are removed during refining of the oil (Dollear, 1969), but the meal would still be toxic and could be used only as fertilizer unless detoxified by some suitable process. Mould growth on oilseeds may also produce undesirable chemical changes in the lipids so that oils extracted from mould-damaged nuts may be of lower quality than might be desired (Eggins & Coursey, 1968).

Another product of importance to the food industry which has been shown to be subject to occasional aflatoxin contamination is soya bean (Shotwell *et al.,* 1969; Bean, Schillinger & Klarman, 1972). If the beans are contaminated the process applied for extracting the oil would not remove any significant amount of toxin and the crude soya flour would still be contaminated. Aflatoxin would not be destroyed during the fermentation process used for production of soy sauce, so that it is essential to use only uncontaminated beans for such processes (Maing, Ayres & Koehler, 1973). This applies also to manufacture of oilseed-based products such as peanut butter, where significant levels of aflatoxin have been reported in some surveys. In 1973, the U.S. FDA reported that 4% of 141 peanut butter samples tested contained > 20 µg/kg aflatoxin and that 25% of the samples contained > 1 µg/kg (Anon., 1973b).

(c) *Fruits and fruit products*

Most fresh fruits are subject to mould spoilage, if stored for excessive periods of time and/or under adverse conditions. The possibility that mould growth could result in the formation of significant quantities of mycotoxin in the fruit and in fruit products has been investigated by a number of workers. The consumer will reject as unacceptable obviously mouldy fruits, but fruits used for processing (e.g. to be made into jams and preserves, fruit pie fillings, fruit juices, etc.) might not be examined so carefully.

Of particular relevance is the commercial pressing of apples, pears, grapes, etc. and the production of fruit pulps and concentrates. Apple juices and ciders are frequently made from fruits which include gleaned 'fallers' and rejects from the fresh-fruit market. *Penicillium expansum,* the mould most frequently implicated as the cause of rot in apples, produces significant amounts of patulin. Wilson & Nuovo (1973) tested 60 isolates of *P. expansum* and found that all produced patulin in apples at levels of 9–146 mg/kg of the expressed juice. These workers also examined 195 samples of fresh apple juice from commercial cider mills in the U.S.A. and detected patulin at levels up to 45 mg/kg in 9 samples. The mills producing patulin-contaminated juices did not separate decayed apples before pressing. These observations confirm those made in Canada by Scott *et al.* (1972). They reported the presence of 1 mg/kg of patulin in one sample of commercial bottled 'cider' (unfermented), but were unable to detect patulin in 11 samples from other sources. Subsequently, Scott (1972) reported the detection of 20–100 µg/kg of patulin in 25 commercial apple juices. Unlike the situation with aflatoxin in malted barley (Campbell, 1972), patulin is destroyed by yeast fermentation of apple juice. Other workers are at present investigating the occurrence of toxins such as citrinin and byssochlamic acid in fruits and fruit products.

3. Mycotoxins in Mould-spoiled Processed Foods

Ample evidence exists that growth of toxinogenic fungi on consumer foods can result in formation of mycotoxins in the food (Frank, 1966; Hanssen & Jung, 1973). With the exception of bread (Hanssen & Jung, 1973) and spaghetti (van Walbeek, Scott & Thatcher, 1968) there are few published examples of mycotoxin contamination of naturally-moulded processed foods. Toxinogenic fungi have been isolated from mouldy foods implicated as the possible cause of incidents of food poisoning (Van Welbeek *et al.*, 1968) and from many other foods including mould-spoiled meat products (Jarvis & Rhodes, unpublished data) which had been subject to excessive refrigerated storage. However, evidence for the presence of specific mycotoxins in these foods has not been obtained.

Products such as marzipan and persipan may contain aflatoxin resulting from mould growth during storage of the blanched almonds, peach and apricot seeds prior to blending with sugar and drying. Hanssen & Jung (1973) detected aflatoxins at up to 2 µg/kg (4 of 16 samples) of marzipan and up to 5 µg/kg (10 of 20 samples) of persipan. However, the contribution of products such as these to our total diet is extremely small.

Inevitably one must pose the question: what is the potential hazard from eating mouldy foods? There is no direct evidence that such foods are a hazard to health, but there is also no evidence that they are not harmful. On the basis that mould spoilage is *a priori* evidence of excessive storage of food products, and that the sale of such products in the U.K. contravenes the *Food and Drugs Act* (1955), it is obviously essential to reject mould-spoiled foods. Cutting or scraping mould growth from the surface of mouldy foods will not directly lessen any hazard since it has been shown by Lie & Marth (1967), Frank (1968*a*), Hanssen & Jung (1973) and other workers that mycotoxins can diffuse some considerable distance into foods.

4. Mycotoxins in Animal Products

Consumption of aflatoxin-contaminated feeds by lactating animals results in excretion of modified aflatoxins (aflatoxins M_1 & M_2) into the milk. The amount of aflatoxin M_1 secreted is directly related to the level of aflatoxin B_1 in the diet of the animal (Purchase, 1973). Although the concentration of aflatoxin M may be reduced in bulk milk by dilution with uncontaminated milk, its presence has been detected in milk and milk products in several countries (Table 3) (Hanssen & Jung, 1973; Kiermeier, 1973a,b; Purchase, 1973; Schuller, Verhülsdonk & Paulsch, 1973).

Aflatoxins B_1, B_2 and M_1 have been detected also in the tissues of pigs (Table 4; Krogh et al., 1973), beef cattle (Purchase, 1973) and poultry (Van Zytveld, Kelley & Dennis, 1970) fed on aflatoxin-contaminated feeds. Earlier

Table 3

Occurrence of aflatoxin M in commercial milk and milk products

Country	Type of milk sample or product	No. of samples tested	No. of samples positive	Level of contamination (μg/kg)	Reference
U.K.	Dried, skim & full-cream	19	0	*	1
U.S.A.	Bulk	?	<1%	0.05–0.4	2
	Cottage cheese	209	15	0.05–0.4	3
	Dried, skim	93	8	0.05–0.4	2
Germany	Single Farms	36	12	0.04–0.25	4
	Bulk, ex Tanker	25	16	0.04–0.13	4
	Dried, Skim	27	17	trace–>2	5
	Dried, Full-cream	28	19	trace–>2	5
S. Africa	Bulk	21	5	trace–>0.16	6

*, not detected by duckling assay therefore < 2 μg/kg in reconstituted sample.

References
1. Allcroft & Carnaghan (1963)
2. Anon (1973b)
3. Brewington, Weihrauch & Ogg (1970)
4. Kiermeier (1973a)
5. Hanssen & Jung (1973)
6. Purchase & Vorster (1968)

Table 4

Occurrence of mycotoxins in tissues of pigs which passed post-mortem veterinary inspection

Toxin	Max. level in feed (μg/kg)	Max. toxin level (μg/kg) in			Reference
		Liver	Kidneys	Muscle, Adipose Tissue	
Aflatoxin $B_1 + B_2$	300	23	20	1	Krogh et al. (1973)
	600	54	53	1	Krogh (1972)
Ochratoxin A	27 500		67	80	Bald & Krogh (1972)

workers (Allcroft & Carnaghan, 1963; Keyl *et al.*, 1970) were unable to detect aflatoxin residues in meat tissues and the recent positive results may reflect the more sensitive analytical procedures now available. However, the levels of occurrence of aflatoxins in animal products is $< 0.1\%$ of the amount of toxin ingested.

The apparent level of aflatoxin M_1 in milks is lowered significantly by processes such as drying, pasteurizing and sterilizing (Table 5; Purchase *et al.*, 1972). It is probable, therefore, that the low levels which may occur in bulk milks will be decreased further by commercial processing. Although there are no published data it is also possible that aflatoxin residues in meat may be decreased to insignificant levels by cooking.

A further example of mycotoxin contamination of animal tissues comes from Danish workers. A disease syndrome of pigs, known for more than 40 years in certain areas of Northern Denmark, has been shown to be caused by feeding mould-damaged cereals containing mycotoxins such as citrinin, and ochratoxin. Contamination of the liver, kidneys, muscle and adipose tissues with ochratoxin has been observed at levels up to 80 μg/kg in meat which would pass veterinary meat inspection (Bald & Krogh, 1972). In more advanced cases the disease is characterized by pathological changes of the kidneys. The disease syndrome (porcine mycotoxic nephropathy) has been observed in Denmark and Ireland (Krogh, 1972) and in Poland (Campbell, pers. comm., 1973). Consequently products made from pork (e.g. bacon, ham and other cured meats) could be contaminated at low levels with these mycotoxins. Studies of the effect of curing and cooking on meat contaminated deliberately with ochratoxin A have demonstrated some instability of the toxin but whether degradation products

Table 5

Effects of processing on aflatoxin M in milk (modified from Purchase et al., 1972)

Process	Conditions	Aflatoxin M (μg/kg)
Freeze-drying	—	385
Pasteurization	62°/30 min	260
	72°/45 sec	210
	80°/45 sec	140
Sterilization (in bottle)	115°/45 sec + 120°/20 min	72
Evaporated milk	40°/45 mmHg pressure	140
Roller drying, of evaporated milk	under reduced pressure at 40°	150
Roller drying, of evaporated milk	Atmospheric pressure	94
Spray drying	180° inlet air; 80° exhaust air	52
Cottage cheese preparation:		
Cheese	made from milk pasteurized at	ND
Whey	80° without addition of rennet	160

ND, not detected.

are also toxic is not known at the present time (Jarvis, unpublished observations).

Whether or not the observed levels of occurrence of toxins such as ochratoxin present a hazard to the consumer is difficult to assess. Certainly the levels of occurrence are much higher than the reported levels of nitrosamines in cured meats. Whether potentiation of toxic effects occurs between toxins such as ochratoxin and nitrosamines, if present together, is also unknown at the present time, but synergism between mycotoxins has been reported previously (Edwards & Wogan, 1968; Lee et al., 1968).

5. Mycotoxins in Mould-ripened Foods

Concern has been expressed from time to time that mould-ripened foods might present a health hazard to the consumer (Hobbs, 1973; Jarvis, 1972). Such foods include mould-ripened cheeses, e.g. Stilton, Danish blue, Roquefort and Camembert, and various meat products, e.g. 'continental sausages' and country-cured hams.

(a) Meat products

Workers in Germany, Poland and the U.S.A. have examined fungal isolates obtained from naturally contaminated, ripened, meat products (Hoffman et al., 1971; Ciegler et al., 1972; Mintzlaff, Ciegler & Leistner, 1972a,b; Strzelecki, 1972, 1973; Strzelecki & Badura, 1972). In one series of experiments, 422 strains of Penicillium spp., isolated from salamis originating in 11 European countries, were tested for ability to produce certain specific mycotoxins on laboratory media (Mintzlaff et al., 1972b). The results are summarized in Table 6. Strains of Penicillium expansum and P. viridicatum constituted > 50%

Table 6

Mycotoxin production by 422 penicillia isolated from commercial salami

Mycotoxins	No. producing strains	No. of strains analysed (%)	Toxin production in YES broth* (mg/100 ml)
Penicillic acid	44	10.4	0.1–50
Ochratoxin A	17	4.0	0.4–64
Tremortin A	11	2.6	0.1–15
Citrinin	10	2.4	0.1–40
Patulin	6	1.4	2.0–48
Total	88	20.9	

* YES, Yeast extract-sucrose broth.
Data from Mintzlaff et al. (1972b).

of the organisms identified by Mintzlaff *et al.* (1972*b*) and together they constituted the largest number of toxinogenic isolates. For *P. viridicatum,* > 80% of the 32 isolates were able to produce one or more toxins on laboratory media, but toxins were not detected when the organisms were grown on meat products under controlled conditions. However, biological tests revealed the presence of some unidentified toxins in a proportion of the test sausages. Evidence for the complexing of penicillic acid by sulphydryl compounds (e.g. cysteine and glutathione) present in meats has been reported (Ciegler *et al.,* 1972). These complexes, and also those formed by patulin (Hoffman *et al.,* 1971) have been found to be non-toxic in biological test systems. It is possible that other mycotoxins may also be complexed by constituents of meat.

Investigations with country-cured hams (Escher, Koehler & Ayres, 1973; Halls & Ayres, 1973) have shown that isolates of *Aspergillus versicolor, A. ochraceus* and *Penicillum viridicatum* from naturally-ripened hams are able to produce sterigmatocystin and ochratoxin both in laboratory media and in ham (Table 7). Production of aflatoxin on ham has been reported by Bullerman & Ayres (1968) and by Strzelecki (1973) but results of the latter worker show > 90% reduction in the aflatoxin content during storage of the ham. However, studies of the incidence of specific mycotoxins in country-cured hams seem not to have been made.

(b) *Cheese*

Strains of *Penicillium roqueforti, P. camemberti* and *P. caseicolum,* which are used as starter organisms for mould-ripened cheeses, have been shown not to produce aflatoxins (Frank, 1968*b*; Bulinski, 1972). Aflatoxins have not been detected in samples of commercial Roquefort, Camembert, Rohpol or 'American' blue cheeses (Shih & Marth, 1969; Bulinski, 1972). However, tests for other mycotoxins in cheese which might be produced by cheese starter

Table 7
Production of mycotoxins in country-cured ham

Organisms	Toxin	Temp/Time	NaCl (%)	Toxin level in ham (µg/kg)
Aspergillus versicolor	Sterigmatocystin	20°/14 days	N.A.	0.16–0.32
		28°/14 days		0.24–0.80
Aspergillus ochraceus	Ochratoxin A	25°/10 days	4.1	5.00–7.90
			5.1	0.77
	Ochratoxin B	25°/10 days	4.1	3.50–4.50
			5.1	0.12

Data from Escher *et al.,* 1973; Halls & Ayres, 1973.

moulds have not been reported. Recently, a West German Newspaper published a report entitled 'Questionable Cheese Delicacies'. The translation of the report (Anon., 1973a) reads:

'Cheese specialities such as the French Roquefort, the Italian Gorgonzola or the English Stilton should preferably not be eaten every day. The basis of their original taste is a fungus poison with the name *Penicillium roqueforti*. When scientists injected rats with minute quantities of 1 or 2 mg of this substance, the animals died within a few hours. Even after injecting 1/10 of this amount, the scientists observed a swelling of the stomach and intestinal tract as well as changes in the lungs, liver, kidney and brain of the animals. Even if experiments on animals are not really transferable to human conditions scientists of the Department of Biochemistry and Plant Pathology of the American University of Wisconsin in Madison advise to be cautious with these cheese varieties.'

This report was based on misrepresentation of facts arising from the work of Wei *et al.* (1973). They showed that a strain of *Penicillium roqueforti* isolated from toxic animal feed, and a culture collection strain, produced a new mycotoxin (PR toxin) with LD_{50} values in mice of 11 and 115 mg/kg by IP and oral routes, respectively. At the present time there is no evidence that this toxin is produced in blue cheeses; unfortunately, there is also no evidence that this or some other mycotoxin is not produced in mould-ripened cheeses. It is clear that misinformed press reports of this kind could be very harmful to trade in particular commodities. Equally it is clear that here is a matter which warrants early investigation. Unfortunately, the results of such an investigation could inadvertently 'lift the lid of Pandora's box', and provide the growing consumer-protection lobby with ammunition for further vilification of the food manufacturing industry. The remedy lies in ensuring that fermentation strains used commercially do not produce mycotoxins in the product.

6. Mycotoxins in Fermentation Products Intended for Food Use

Microfungi are used commercially for the production of fungal biomass and for the preparation of metabolites such as enzymes and acids which are intended for food use. Many mycotoxins are known now to be antibiotics which were rejected by the pharmaceutical industry as being too toxic for clinical use (e.g. patulin and citrinin). Before being accepted for food use it is essential that fungal products are shown to be atoxic. In the U.S.A. the Federal Regulations for specific products require that they be shown to be free from aflatoxin. Campbell (pers. comm., 1973) has recently stated: 'the F.D.A. requires assurances that products are free from aflatoxins . . . before issuing a food additive regulation. So far the emphasis has been on aflatoxin but other mycotoxins are also considered. . .'. This applies equally when the metabolites are of bacterial origin since it is argued that the fermentation substrates might have been contaminated with mycotoxins (e.g. it might be argued that aflatoxin M could be present in the skim milk used in the production of nisin). Tests on enzyme preparations and other food additives from a number of microfungi have failed to reveal any detectable levels of aflatoxin or other specific mycotoxin (Jarvis, unpublished data).

Mycotoxin problems associated with biomass production may be potentially more acute than with fungal metabolites since there will be less purification and many mycotoxins are associated largely with the fungal mycelium. Mutation of organisms, slight variations in fermentation conditions, etc. might be sufficient to stimulate a change in metabolism and lead to formation of mycotoxins. However, careful control in the selection and use of organisms should minimize any hazards from fungal biomass.

7. Control of Mycotoxins in Foods

Many countries (including those in the EEC) already have, or are promulgating, regulations to control the importation of aflatoxin-contaminated commodities. As yet regulations for other mycotoxins have not appeared. Suggested guidelines for aflatoxin levels in imported and home-produced foods in a number of countries are given by Arrhenius (1973). In the U.S.A. interpretation (or misinterpretation?) of the Delaney Ammendment leads to the situation where it is illegal to sell a food containing any detectable quantity of a mycotoxin (Goldblatt, 1973). By contrast, the United Nations Protein Advisory Group accepts a maximum level of 30 μg of aflatoxin/kg in food supplements for under-nourished children in developing countries. Undoubtedly the argument used is that it is better to die of cancer in old age than from malnutrition in childhood.

If it is accepted that certain food commodities may from time to time be contaminated with mycotoxins, how may the hazard be reduced to a minimum? Hand selection or automatic sorting machines are used widely for separating damaged or mouldy peanuts from undamaged nuts. Similar methods may be used for other commodities. Numerous methods have been tested for destroying aflatoxin in agricultural produce. Although aflatoxins are relatively stable to heat (Dollear, 1969) recent studies with cottonseed meal and cereal grains demonstrate the feasibility of decontamination by very high temperature treatment in the presence of ammonia (Campbell, 1972; Goldblatt, 1973). The ammonia-treated product may subsequently be used for animal feed. Treatment of agricultural products with ethylene oxide (Mayr, 1973) not only kills insects and fungal contaminants but also destroys aflatoxin. This mode of decontamination warrants further investigation, although problems with toxic reaction products of the ethylene oxide could make the method unacceptable. Studies of the thermal stability of ochratoxin in cereals (Trenk, Butz & Chu, 1971) and in meat (Jarvis, unpublished results) demonstrate some instability but whether the degradation products are also toxic is not known at the present time. In general, destruction of mycotoxins in food products warrants further investigation, especially with the present world shortages of protein foods, but the onus in control measures must lie with improved methods for post-harvest storage of food commodities.

8. Conclusion

This review would be incomplete without mention of the situation in developing countries. Many workers have demonstrated an association between consumption of aflatoxin (or other mycotoxin), the nutritional status of the community and the incidence of hepatoma and other disease syndromes in people from developing countries (Barnes, 1970; Crawford, 1971; Shank, 1971). A direct consequence of the introduction of acceptance standards for imported food commodities in the developed countries is that the population of developing countries may be placed at even greater risk than previously. Rejection of mycotoxin-contaminated produce by the Western world may affect the economic viability of the developing countries and may impose a situation where only the lowest grade materials are consumed locally.

Even though we may not be faced with regular consumption of significant quantities of mycotoxins in our diet, we should not ignore the possible consequences of occasional consumption of mycotoxins in our food nor forget that some consumers may be at greater risk than the majority. The 2 groups potentially at highest risk are the immigrant population, who may attempt to retain in part their traditional food customs in an alien climate, and the vegetarian or health food addict. The 'pure food lobby' may ensure that their food comes from sources where neither 'chemical' residues of fertilizers, pesticides or fungicides nor 'chemical food additives' occur in their food; yet without pesticides and fungicides, foods stored under *natural* (but not necessarily *good*) conditions may become heavily contaminated with toxinogenic moulds.

The food manufacturing industry in the U.K. and elsewhere is aware of the need to ensure that its raw materials are free from aflatoxins, but attention to other mycotoxins is warranted in certain foodstuffs. In a society which is becoming increasingly oriented towards consumer protection, and with the need to consider the introduction of alternatives to many traditionally used raw materials, the food industry must continue to ensure that its products are wholesome.

Whilst, at one time, it was believed that mycotoxins would present a potential hazard only in plant products, evidence is now accumulating on the occurrence of mycotoxins in other types of foods. A major difficulty in evaluating the potential hazard to the consumer in the developed countries is a lack of data on the incidence of mycotoxins in many food products. At the present time the available evidence suggests that mycotoxins do not constitute a major health hazard to the majority of consumers. One must, however, remain aware of the potential hazard from mycotoxins. In 1972, Fischbach stated (Fischbach & Rodricks, 1972): 'Mycotoxins may be one of the most significant pollutants known to man and may be the cause of many diseases in man of unknown aetiology. The FDA will err on the side of ensuring safety for the consumer'.

9. Acknowledgements

I am indebted to the Microbiology Panel of the Food R.A. for financing this review and to my colleagues for their constructive comments on the manuscript. Thanks are due also to my wife for her forbearance and for her assistance with the bibliography.

10. References

ALLCROFT, R. & CARNAGHAN, R. B. A. (1963). Groundnut toxicity: an examination for toxin in human food products from animals fed toxic groundnut meal. *Vet. Record* **75**, 259.

ANON (1973*a*). Bedenkliche Käse-Delikatesse. *Rhein-Zeitung* 22nd August 1973.

ANON (1973*b*). *Food Chemical News* 3rd December 1973, p. 49.

ARRHENIUS, E. (1973). Mycotoxicosis; An old health hazard with new dimensions. *Ambio* **2**, 49.

BALD, P. & KROGH, P. (1972). Ochratoxin residues in bacon pigs. Abstracts I.U.P.A.C. Symposium *Control of Mycotoxins*. Göteborg, Sweden.

BARNES, J. M. (1970). Aflatoxin as a health hazard. *J. appl. Bact.* **33**, 285.

BEAN, G. A., SCHILLINGER, J. A. & KLARMAN, W. L. (1972). Occurrence of aflatoxins and aflatoxin-producing strains of *Aspergillus* spp. in soybeans. *Appl. Microbiol.* **24**, 437.

BECROFT, D. M. O. & WEBSTER, D. R. (1972). Aflatoxins and Reye's Disease. *Br. med. J.* **3**, 117.

BREWINGTON, C. R., WEIHRAUCH, J. L. & OGG, C. L. (1970). Survey of commercial milk samples for aflatoxin M. *J. Dairy Sci.* **53**, 1509.

BULIŃSKI, R. (1972). Examination for aflatoxins of moulds used in cheese making and of mould ripened cheeses. *Roczn. Inst. Przcm. Mlecz.* **14**, 35. (*Dairy Sci. Abs.* **34**, no. 5133).

BULLERMAN, L. B. & AYRES, J. C. (1968). Aflatoxin-producing potential of fungi isolated from cured and aged meats. *Appl. Microbiol.* **16**, 1945.

CAMPBELL, A. D. (1972). Chemical and biological alterations of mycotoxins and their relationship to control of the mycotoxin problem. Abstracts I.U.P.A.C. Symposium *Control of Mycotoxins,* p. 6; Göteborg, Sweden.

CAMPBELL, T. C. & SALAMAT, L. (1971). Aflatoxin ingestion and excretion by humans. In *Mycotoxins in Human Health.* Ed. I. F. H. Purchase. London: Macmillan.

CIEGLER, A., MINTZLAFF, H. J., WEISLEDER, D. & LEISTNER, L. (1972). Potential production and detoxification of penicillic acid in mold-fermented sausage (salami). *Appl. Microbiol.* **24**, 114.

CRAWFORD, M. A. (1971). Epidemiological interactions. In *Mycotoxins in Human Health.* Ed. I. F. H. Purchase. London: Macmillan.

DOLLEAR, F. G. (1969). Detoxification of aflatoxins in foods and feeds. In *Aflatoxin.* Ed. L. A. Goldblatt. New York: Academic Press.

EDWARDS, G. S. & WOGAN, G. N. (1968). Acute and chronic toxicity of rubratoxin in rats. *Fedn Proc. Fedn Am. Socs exp. Biol.* **27**, 552.

EGGINS, H. O. W. & COURSEY, D. G. (1968). The industrial significance of the biodeterioration of oilseeds. *Int. Biodetn Bull.* **4**, 29.

ESCHER, F. E., KOEHLER, P. E. & AYRES, J. C. (1973). Production of Ochratoxins A and B on Country Cured Ham. *Appl. Microbiol.* **26**, 27.

FISCHBACH, H. & RODRICKS, J. V. (1972). Current efforts of the U.S. Food and Drug Administration to control mycotoxins in foods. I.U.P.A.C. Symposium *Control of Mycotoxins.* Göteborg, Sweden.

FOOD & DRUGS ACT (1955). London: H.M.S.O.

FORGACS, J. (1965). Stachybotryotoxicosis and moldy corn toxicosis in *Mycotoxins in Foodstuffs.* Ed. G. N. Wogan. Cambridge, Mass: M.I.T. Press.

FRANK, H. K. (1966). Aflatoxine in Lebensmitteln. *Archiv Lebensmittelhyg.* **17**, 237.

FRANK, H. K. (1968*a*). Diffusion of aflatoxin in foodstuffs. *J. Fd Sci.* **33**, 98.

FRANK, H. K. (1968*b*). Sind Mykotoxine in unserer Nahrung eine Gefahr? *Therapiewoche* **29**, 1172.

GOLDBLATT, L. A. (1973). Learning to live with Mycotoxins: Aflatoxin – a case history. *Pure appl. Chem.* **35**, 223.

HALLS, N. A. & AYRES, J. C. (1973). Potential production of sterigmatocystin on country-cured ham. *Appl. Microbiol.* **26**, 636.

HANSSEN, E. & JUNG, M. (1973). Control of aflatoxins in the food industry. *Pure appl. Chem.* **35**, 239.

HESSELTINE, C. W. (1973). Recent research for the control of mycotoxins in cereal. *Pure appl. Chem.* **35**, 251.

HOBBS, B. C. (1973). Bioassay methods for mycotoxins. In *The Microbiological Safety of Foods.* Eds B. C. Hobbs & J. H. Christian. London: Academic Press.

HOFFMAN, K., MINTZLAFF, H. J., ALPERDEN, I. & LEISTNER, L. (1971). Untersuchung über die Inaktivierung des Mykotoxins Patulin durch Sulfhydrylgruppen. *Fleischwirtschaft,* **51**, 1534, 1539.

JARVIS, B. (1972). The significance of microbial toxins in foods with respect to the health of the consumer. In *Health and Food.* Eds G. G. Birch, L. F. Green & L. G. Plaskett. London: Applied Science Publishers.

JEMALI, M. & LAFONT, P. (1972). Das verhalten des Aflatoxin B$_1$ im Verlauf der Brotbereitung. *Getreide Mehl Brot* **26**, 193.

JOFFE, A. Z. (1971). Alimentary toxic aleukia. In *Microbial Toxins,* Vol. VII. Eds S. Kadis, A. Ciegler & S. J. Ajl. New York: Academic Press.

KEYL, A. C., BOOTH, A. N., MASRI, M. S., GUMBMANN, M. R. & GAGNE, W. E. (1970). Chronic effects of aflatoxin in farm animal feeding studies. In *Toxic Micro-organisms.* Ed. M. Herzberg. Washington D.C.: U.S. Govt. Printing Office.

KIERMEIER, F. (1973*a*). Aflatoxic residues in fluid milk. *Pure appl. Chem.* **35**, 271.

KIERMEIER, F. (1973*b*). Mykotoxine in Milch und Milchprodukten. *Z. Lebensmittelunters. u-Forsch.* **151**, 237.

KROGH, P. (1972). Nephropathy caused by mycotoxins from *Penicillium* and *Aspergillus.* *J. gen. Microbiol.* **73**, xxxiv.

KROGH, P., HALD, B., HASSELAGER, E., MADSEN, A., MORTENSEN, H. P., LARSEN, A. E. & CAMPELL, A. D. (1973). Aflatoxin residues in bacon pigs. *Pure appl. Chem.* **35**, 275.

LEE, D. J., WALES, J. H., AYRES, J. L. & SINNHUBER, R. O. (1968). Synergism between cyclopropenoid fatty acids and chemical carcinogens in rainbow trout. *Cancer Res.* **28**, 2312.

LIE, J. L. & MARTH, E. H. (1967). Formation of aflatoxin in Chedder cheese by *Aspergillus flavus* and *Aspergillus parasiticus. J. Dairy Sci.* **50,** 1708.

MAING, I. Y., AYRES, J. C. & KOEHLER, P. E. (1973). Persistence of aflatoxin during the fermentation of soy sauce. *Appl. Microbiol.* **25,** 1015.

MAYR, G. (1973). Untersuchungen über die Wirkung von Äthylenoxid zur Beseitigung von Schimmelpilzen und deren Stoffwechsel-produkten. Doctoral Thesis. Dept. Fd Technol. Rheinischen Friedrich-Wilhelms-Universität zu Bonn, Germany.

MINTZLAFF, H.-J., CIEGLER, A. & LEISTNER, L. (1972a). Pathogene und Toxinogene Hefen und Schimmelpilze in Fleisch und Fleischwaren. *Archiv Lebensmittelhyg.* **12,** 286.

MINTZLAFF, H.-J., CIEGLER, A. & LEISTNER, L. (1972b). Potential mycotoxin problems in mould-fermented sausage. *Z. Lebensmittelunters u.-Forsch.* **150,** 133.

PERONE, V. B., SCHEEL, L. D. & MEITUS, R. J. (1964). A bioassay for the quantification of cutaneous reactions associated with pink-rot celery. *J. Invest. Dermatol.* **42,** 267.

PURCHASE, I. F. H. (1971). *Mycotoxins in Human Health.* London: Macmillan.

PURCHASE, I. F. H. (1973). The control of aflatoxin residues in food of animal origin. *Pure. appl. Chem.* **35,** 283.

PURCHASE, I. F. H., STEYN, M., RINSMA, R. & TUSTIN, R. C. (1972). Reduction of the aflatoxin M content of milk by processing. *Fd Cosmet. Toxicol.* **10,** 383.

PURCHASE, I. F. H. & VORSTER, L. J. (1968). Aflatoxin in commercial milk samples. *S. African Med. J.* **42,** 219.

RAHNEMA, R. (1972). Control of mycotoxins in pistacchio nuts. Abstracts, I.U.P.A.C. Symposium *Control of Mycotoxins.* Göteborg, Sweden.

SAITO, M., ENOMOTO, M., TATSUNO, T. & URAGUCHI, K. (1971). Yellowed rice toxins: luteoskyrin and related compounds, chlorine-containing compounds and citrinin. In *Microbial Toxins,* Vol. VI. Eds A. Ciegler, S. Kadis & S. J. Ajl. New York: Academic Press.

SCHULLER, P. L., VERHÜLSDONK, C. A. H. & PAULSCH, W. E. (1973). Analysis of aflatoxin M₁ in liquid and powdered milk. *Pure appl. Chem.* **35,** 291.

SCOTT, P. M. (1972). Occurrence of ochratoxin A and citrinin in cereals and patulin in apple juice. Abstracts, I.U.P.A.C. Symposium *Control of Mycotoxins,* Göteborg, Sweden.

SCOTT, P. M. (1973). Mycotoxins in stored grain, feeds and other cereal products. In *Grain Storage – Part of a System.* Eds R. N. Sinha & W. E. Muir. Westport, U.S.A.: Avi Publishing Co.

SCOTT, P. M., MILES, W. F., TOFT, P. & DUBÉ, J. G. (1972). Occurrence of patulin in apple juice. *J. agr. Fd Chem.* **20,** 450.

SHANK, R. C. (1971). Dietary aflatoxin loads and the incidence of human hepatocellular carcinoma in Thailand. In *Mycotoxins in Human Health.* Ed. I. F. H. Purchase. London: Macmillan.

SHIH, C. N. & MARTH, E. H. (1969). Aflatoxins not found in commercial mould ripened cheeses. *J. Dairy Sci.* **52,** 1681.

SHOTWELL, O. L., HESSELTINE, C. W., BURMEISTER, H. R., KWOLEK, W. F., SHANNON, G. M. & HALL, H. H. (1969). Survey of cereal grains and soybeans for the presence of aflatoxin. II. Corn & Soybeans. *Cereal Chem.* **46,** 454.

STRZELECKI, E. L. (1972). Extraction and determination of *Aspergillus flavus* metabolites – aflatoxins – from meat products. *Acta microbiol. pol. Ser. B.* **4,** 155.

STRZELECKI, E. L. (1973). Behaviour of aflatoxins in some meat products. *Acta microbiol. pol. Ser. B.* **5,** 171.

STRZELECKI, E. L. & BADURA, L. (1972). Occurrence of aflatoxinigenic molds on "dry Cracower sausage". *Acta microbiol. pol. Ser. B.* **4,** 233.

TRENK, H. L., BUTZ, M. E. & CHU, F. S. (1971). Production of ochratoxins in different cereal products by *Aspergillus ochraceus. Appl. Microbiol.* **21,** 1032.

UENO, Y. (1971). Toxicological and biological properties of Fusarenan-X, a cytotoxic mycotoxin of *Fusarium nivale.* In *Mycotoxins in Human Health.* Ed. I. F. H. Purchase. London: Macmillan.

VAN WALBEEK, W., SCOTT, P. M. & THATCHER, F. S. (1968). Mycotoxins from food-borne fungi. *Can. J. Microbiol.* **14**, 131.

VAN ZYTVELD, W. A., KELLEY, L. J. & DENNIS, S. M. (1970). Aflatoxins: The presence of aflatoxins or their metabolites in livers and skeletal muscles of chickens. *Poult. Sci.* **49**, 1350.

WEI, R. D., STILL, P. E., SMALLEY, E. B., SCHNOES, H. K. & STRONG, F. M. (1973). Isolation and partial characterization of a mycotoxin from *Penicillium roqueforti.* *Appl. Microbiol.* **25**, 111.

WILSON, D. M. & NUOVO, G. J. (1973). Patulin production in apples decayed by *Penicillium expansum. Appl. Microbiol.* **26**, 125.

Subject Index

Acetate levels, ruminal, 113

Acridine orange, treatment of plasmids with, 227, 230

Actinomycetes, spores of in soils, 26
taxonomy and classification of, 20

Additives, silage, 116, 117

Aeromonads, as cause of death in fish, 59

Aeromonas hydrophila, effects of on fish, 59

Aero. salmonicida, attempts at vaccine production against, 59
cause of furunculosis by, 56, 59

Aflatoxin, 243, 244, 245, 248

Aggregate, soil, partially anaerobic, in an aerobic environment, 28

Aggregates, soil, methods for microbiological analysis of, 30, 31
role of micro-organisms in formation of, 32, 33

Alimentary toxicaleukia, 248

Alkanes, anaerobic dissimilation of, 99
degradation of, 98, 105

Amino acids, deamination of, 130

Ammonium, effect of on fungal growth on root exudates, 45

Anaerobic starch fermentation plate, 184

Animal feeds, salmonellae in, 169, 170

Anthrax bacilli, in bone meal, 171

Apples, patulin produced in, 255

Aromatic fraction of oils, degradation of, 97

Arthrobacter, production of cytokinins by, 42

Artificial rumen, 127, 128, 136, 137

AS–2 viruses, 77

Ascorbic acid, influence of on anti-microbial effects of nitrite, 6

Aspergillus flavus, aflatoxinogenic strains of, 254
strains of found in home-grown cereals, 244

Auxin, effect of on root growth, 41, 42

Azotobacter chroococcum, bacterization with, 46
effect of on roots, 42, 43

Bacillus anthracis, means of distinguishing from *B. cereus*, 207

B. subtilis, disease control by, 47

B. thuringiensis, pathological symptoms produced by, 208, 209

Bacteria, effects of on nutrient uptake by roots, 40

Bacterial count, in poultry carcasses, 154, 155

Bacteriophages, resemblance of LPP viruses to, 64

Bacterization, 45, 46, 47

Barley, brewing, mycotoxin in, 254

BC medium, for *Bacillus cereus*, 199

Biological control, naturally occurring, 48

Biomass, conversion of carbon content of alkanes to, 98
mycotoxins associated with production of, 261, 262

Biomass carbon, calculations of, 24

Biovolume, microbial, in soils, estimate of, 22, 23, 24

Blood agar, used for isolation of *Bacillus cereus*, 198, 199, 203

Botulism, outbreaks of caused by home-cured meats, 1

Bread, made from aflatoxin-contaminated cereal, 254

Broadbalk field, non-symbiotic nitrogen fixation in, 30

Broiler house litter, microbial flora of, 157, 158

Calves, *E. coli* diarrhoea in, 228, 229, 230, 238, 240

Canned cured meat, a nitrite-derived inhibitor present in, 13

Canned cured meats, stability of, 7

Cellulose, breakdown of in rumen, 128

Cellulolytic bacteria, growth of, 134
isolation of, 128, 132

Cellvibrio, algal-lysing material produced by, 72

Cheese, mould-ripened, mycotoxins in, 260, 261

Chicken muscle, effect of psychrophilic bacteria on, 155

Chocolate, connection with *Salm. eastbourne* outbreak, 17

Chloramphenicol resistance, in *Salm. typhi*, 172

Chorella pyrenoidosa, viruses of, 71

Cider, patulin levels in, 255, 256, 258

Ciliate protozoa, in rumen, 130, 138

Citrinin, toxicity of, 261

Clostridia, silage fermented by, 111, 116

Clostridium spores, effect of nitrite on germination of, 14